MICROSTRUCTURAL DEVELOPMENT AND STABILITY IN HIGH CHROMIUM FERRITIC POWER PLANT STEELS

MICROSTRUCTURAL DEVELOPMENT AND STABILITY IN HIGH CHROMIUM FERRITIC POWER PLANT STEELS

Edited by
A. Strang & D. J. Gooch

MICROSTRUCTURE OF HIGH TEMPERATURE MATERIALS
Number 1

Series Editor
A. Strang

THE INSTITUTE OF MATERIALS

Book 667
Published in 1997 by
The Institute of Materials
1 Carlton House Terrace
London SW1Y 5DB

ISSN 1366–5510
ISBN 1 86125 021 5

Typeset by
Fakenham Photosetting Ltd
Fakenham, UK

Printed and bound by
The University Press
Cambridge, UK

Contents

Preface

Materials used for high temperature components in plant applications, such as the power generation, aerospace and chemical industries, are known to undergo microstructural changes in service which affect their mechanical properties. These take the form of particle coarsening, phase dissolution and recovery effects, leading to material softening and reductions in tensile and creep strengths. Precipitation of new phases and diffusion of impurities to grain boundaries can also result in reduced creep ductility, impact strength and toughness. Material degradation processes such as these can compromise the integrity of critical high temperature plant components and lead to reduced operational life and premature failure in service.

These microstructural changes are influenced by the operational conditions experienced by the material in service, and are dependent on the temperature and duration of exposure, as well as any deformation which occurs due to the imposed service stresses. Microstructural degradation also occurs at different rates in different materials and is strongly influenced by material composition, original heat treatment and initial microstructure. An understanding of the physical processes responsible for microstructural degradation, and the mechanisms whereby mechanical properties are changed during service, are important in determining the safe operating life of critical high temperature plant components. Additionally, such an understanding may contribute to the development of new alloy systems with improved long term high temperature properties, thus enabling the development of improved of improved high efficiency generating plant.

This series of specialist conferences is intended to focus attention on the microstructural changes which occur in service exposed materials and, by doing so, on identification of the mechanisms and processes leading to the observed reductions in their mechanical properties. The series will highlight the work currently in progress on the development of improved materials which will be more resistant to microstructural degradation in service. It is intended that the papers presented at each of these conferences will be published in the form of a series of linked Conference Proceedings which will serve to highlight the current state of knowledge in this important area of materials technology.

These are the proceedings of the first conference in the series which was held at Robinson College, Cambridge in June 1996 and which dealt with the microstructural stability of high temperature, creep resistant, 9–12%Cr martensitic power plant steels. The conference attracted 75 delegates from 12 countries.

Series Editor

Andrew Strang
Chairman High Temperature Materials Performance Committee

Future Conferences Planned in the Series

The next conference is planned to take place at Sheffield Hallam University from 24–26 March 1997 and is entitled:

Microstructural Stability of Creep Resistant Alloys for High Temperature Plant Applications

FUTURE CONFERENCES

Microstructure of Advanced Ceramics and Ceramic Systems for High Temperature Applications

Microstructure of Welded Materials and Structures for High Temperature Plant Applications

Microstructural Control and Stability of High Temperature Gas Turbine Blading Alloys

Microstructure of Advanced High Temperature Titanium Alloys

Microstructure of Advanced High Temperature Composites

Microstructural Stability of Corrosion Resistant Coatings for High Temperature Gas Turbine Blading and Combustion Path Components

Microstructural Modelling and High Temperature Property Prediction of Creep Resistant Materials

Quantitative Evaluation of Microstructure for Modelling and Prediction of High Temperature Creep Resistant Materials

Historical Development and Microstructure of High Chromium Ferritic Steels for High Temperature Applications

F.B. PICKERING
Sheffield Hallam University

ABSTRACT

The paper presents an overview of the microstructural evolution in 12%Cr steels for power plant applications. It deals with the compositional factors affecting the control of the constitution of the steels, and with the martensitic and re-austenitisation transformations. A detailed description is given of the tempering characteristics and the associated carbide precipitations and transformations. The effects of alloying on the carbide precipitation with a view to increasing the tempering resistance and high temperature properties is presented. The precipitation of intermetallic compounds to gain extra tempering resistance and high temperature strength is discussed. A description is presented of the phenomena and mechanisms affecting microstructural stability and degradation at high tempering temperatures or in prolonged service, which is followed by a consideration of various embrittling phenomena which may occur especially in the highly alloyed 12%Cr steels. There is a brief account of structure-property relationships. Throughout the paper the relation of the metallurgical phenomena involved in the design of the steels has been discussed.

INTRODUCTION

The high Cr ferritic steels used for power plant[1] are based on a low C (0.10/0.20%) 12%Cr composition. Above about 10%Cr, there is the required level of oxidation resistance for the temperatures involved. In order to increase the efficiency of power plant, higher creep strengths and particularly higher operating temperatures are sought and thus the Cr content may be slightly lowered to 10–12% and the tempering resistance increased by additions of such elements as Mo, W, V, Nb, Ta, etc. Due to the higher operating temperatures, the need to minimise microstructural degradation is required, which means that the carbide precipitates formed during tempering must resist growth, thereby providing microstructural stability. Also, the use of increased N content to increase the overall strength level is quite common. Apart from high creep strength, the steels must also show an adequate level of

creep ductility. In addition, there are room temperature property requirements to be met in terms of an acceptable strength level but more particularly
good toughness. Moreover, the microstructural stability necessary for long
term creep strength must not result in embrittlement effects during the microstructural degradation inevitable during service.

All the above requirements lead to complex interactions between steel composition and constitution, transformation behaviour and carbide precipitation
during heat treatment and service. Moreover, as attempts are made to improve
the high temperature properties by improving the tempering resistance or by
introducing precipitation hardening reactions, often involving intermetallic
compounds, these interactions become even more complicated and difficult to
control. It is the purpose of this review to discuss the various metallurgical
phenomena involved and to attempt to rationalise the interacting effects.

THE CONSTITUTION OF 12%Cr STEELS

Above about 12%Cr suffices to close the γ loop in the Fe–Cr equilibrium diagram but the γ loop is extended appreciably by C, N, Ni and other austenitising elements.[2-4] Equally ferrite forming elements such as Mo, W, V, etc.
which are used to impart tempering resistance, tend markedly to restrict the γ
loop.[5] In simple 12%Cr steels containing 0.1%C and low N, austenite is
stable between ~950°C and 1150°C, Fig. 1, in which temperature range the
steels are usually austenitised. The alloyed 12%Cr steels have therefore a distinct tendency to contain δ ferrite which is detrimental as it detracts from the
potential strength of the steel. In addition, δ ferrite, particularly if present as
films between the austenite grains, which have subsequently transformed to
martensite which has been tempered, can materially lower the toughness.
Also, because of alloy partitioning between the γ and δ phases, the composition of the δ can be so modified that considerable changes occur in the precipitation effects in the δ ferrite.

Because it is necessary to alloy with predominantly ferrite forming elements
in order to increase the tempering resistance, the formation of δ ferrite must
be prevented by the use of an appropriate amount of an austenite forming element, usually Ni. The balancing of the constitution to provide 100% austenite at the austenitising temperature can be determined quantitatively using the
following data[6] derived specifically for 12%Cr steels:

Element	Change in δ ferrite content (%) per mass % alloy addition
N	−220
C	−210
Ni	−20

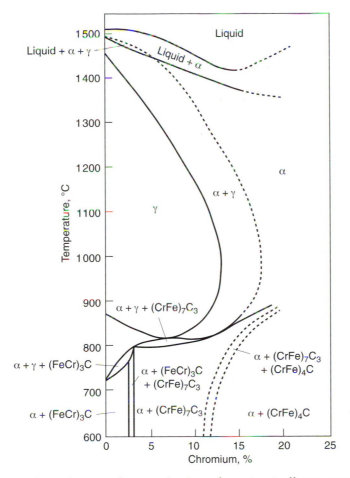

Fig. 1   Effect of Cr on the constitution of Fe–Cr–C alloys containing 0.1%C.
(Note – Cr_4C is the $M_{23}C_6$ carbide)

Element	Change in δ ferrite content (%) per mass % alloy addition
Co	−7
Cu	−7
Mn	−6
W	+3
Mo	+5
Si	+6
Cr	+14
V	+18
Al	+54

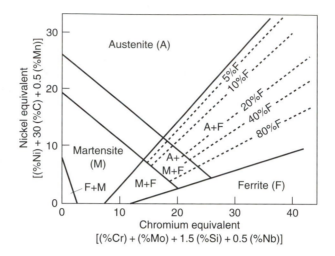

Fig. 2 Schaeffler-Schneider diagram.

Alloying elements such as Nb, Ti and Ta can cause difficulties as they have variable effects, being ferrite formers, but also removing the austenite formers C and N as insoluble carbides/nitrides.[1] As the constitution is determined by the amount of dissolved alloying element, these concentrations may be calculated from appropriate solubility data for nitrides and carbides of Nb, Ti and Ta. But highly reliable data for 12%Cr base compositions is sparse. A first approximation to the constitution can be arrived at by using Cr and Ni equivalent compositions together with the well known Schaeffler diagram modified by Schneider.[7] Various Cr and Ni equivalent equations have been suggested, the following being reasonably appropriate:

$$\text{Cr equivalent} = \text{Cr} + 2\text{Si} + 1.5\text{Mo} + 5\text{V} + 5.5\text{A1} + 1.75\text{Nb} \qquad (1)$$
$$+ 1.5\text{Ti} + 0.75\text{W}$$

$$\text{Ni equivalent} = \text{Ni} + \text{Co} + 0.5\text{Mn} + 0.3\text{Cu} + 25\text{N} + 30\text{C} \qquad (2)$$

The amounts of the alloying elements are in mass %.

A typical Schaeffler-Schneider diagram is shown in Fig. 2, but care should be taken in using the indication as to whether the austenite transforms to martensite because the coefficients for the elements for the Cr and Ni equivalents are not the same as those for their effects on the M_s and M_f temperatures. The cheapest austenite former for balancing the constitution would be C, but this is unacceptable as it would lower the toughness and weldability, impair corrosion resistance and require higher austenitising temperatures to dissolve the carbides, particularly of W, V, Nb, Ti and Ta, which would lead to coarser austenite grain sizes, lower toughness and decreased creep ductility. N can be used, if it is present for increased strength, but the amount required could be excessive. Consequently, Ni is most often used and whilst being less potent than C and N it has fewer detrimental effects. It does, however, slightly decrease the tempering resistance but not excessively and, as will be seen later, its use is also limited as it lowers the Ac_1 temperature. Although disadvantageous economically, Co as an austenite former has the advantage that it does not depress the M_s–M_f temperatures.

THE TRANSFORMATIONS IN 12%Cr STEELS

It is essential, in order to achieve the optimum strength, for the austenite present at the austenitising temperature to transform fully to martensite on cooling to room temperature.[8] Thus the M_f should be above room temperature. For a 0.1%C, 12%Cr steel the M_s is about 300°C, with the M_f being in the range 100–150°C. Alloying used to improve the tempering resistance or to balance the constitution will lower both M_s and M_f temperatures with the tendency for retained austenite to be present if the M_f is close to or below room temperature.[6–9] This is particularly the case during the fairly slow cooling of large section sizes in which some stabilisation of the austenite may occur. Such retained austenite is generally very undesirable because distortion can occur when it transforms, it gives lower strength prior to tempering and after tempering it can lead to untempered martensite because carbide precipitation will have raised the M_f of the retained austenite to above room temperature. Some retained austenite can however be advantageous to toughness as is designed into the 12%Cr, 4%Ni steels by tempering above Ac_1;[10] but these steels are not creep resisting.

Thus the composition of the steel must not only control the constitution, but also control the M_s-M_f temperatures to above room temperature. Some effects of alloying elements on the M_s temperature, specifically with respect to 12%Cr steels are:[6,8]

Element	Change in M_s (°C) per mass %
C	−474
Mn	−33
Ni	−17
Cr	−17
Mo	−21
W	−11
V	−11
Cu	−10
Nb	−11
Si	−11
Co	+15

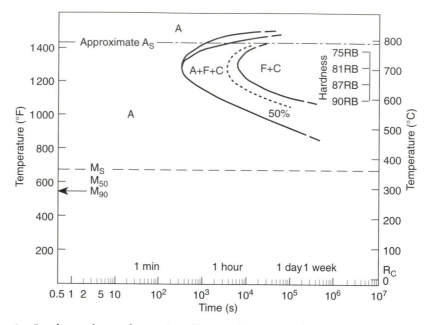

Fig. 3   Isothermal transformation diagram for a 0.1%C, 12%Cr steel.

Many empirical equations allowing calculation of the M_s temperature have been published,[11–13] a typical one being:[8,14,15]

$$M_s\,(°C) = 550°C - 450C - 33Mn - 20Cr - 17Ni - 10W - \qquad (3)$$
$$20V - 10Cu - 11Nb - 11Si + 15Co$$

Such equations really only give a general indication of the M_s, and more recent approaches have been based on thermodynamics but seem to be but little more accurate. The interesting element is Co which, whilst being an austenite former and thus capable of helping to balance the constitution, actually raises the M_s temperature. It therefore can be useful in steels containing large amounts of ferrite formers.[16,17]

In general the continuous cooling or isothermal transformation characteristics of the 12%Cr steels cause no problems. They are air hardenable,[5] forming martensite in the largest sections because the pearlite transformation is greatly retarded and no bainite forms within very extended time periods, Fig. 3. In fact the steels are hypereutectoid, showing pro-eutectoid carbide precipitation of $Cr_7C_3/M_{23}C_6$. The martensite formed is typical low carbon lath martensite. Apart from conditions of toughness, previously mentioned, it is essential that austenite does not form during tempering, i.e. the Ac_1 temperature should be above the maximum tempering temperature employed.[5–8] Ni especially, Mn also and Co to a small extent all lower the Ac_1 temperature. As Ni is the major austenite former used to balance the constitution, too much

Ni can adversely affect the maximum tempering temperature which can be used. For example, a 12%Cr, 4%Ni steel may have an Ac_1 as low as 550°C, and such a low Ac_1 would not enable a sufficiently high tempering temperature for microstructural stability to be achieved and would render the steel unsuitable for use at the required higher operating temperatures. Some effects of alloying elements on the Ac_1 temperature of 12%Cr steels are:[6,8]

Element	Change in Ac_1 (°C) per mass %
Ni	−30
Mn	−25
Co	−5
Si	+25
Al	+30
Mo	+25
V	+50

It can be seen that no more than ~3%Ni can be accommodated if the steels are to be tempered at 650°C, but if reaustenitisation should not occur below 700°C somewhat less than 2%Ni is the limit.

As mentioned previously, a 12%Cr, 4%Ni steel can be used if superior toughness is required. Such a steel would not be suitable for power plant, but some interesting metallurgical effects are shown by this steel which are worthy of mention. If tempering is carried out just above Ac_1 at 600/620°C, retained austenite is produced.[18] This austenite nucleates at tempered martensite lath boundaries and forms an inter-leaved lamellar structure of retained austenite films and tempered martensite laths. Because of the small amount of austenite formed just above Ac_1, equilibrium dictates that the austenite forming elements, particularly, C, N, Mn, Ni, partition heavily to it so that its M_s temperature is below room temperature so that on cooling the austenite is retained. This duplex structure is tough,[19] Charpy values of >100 J at − 40°C being produced due to the crack stopping propensity of the retained austenite. On tempering at higher temperatures, say 680°C, much more austenite is formed on tempering which equilibrium dictates will suffer very much less austenite forming element partitioning. The M_s and M_f temperatures are now well above room temperature so that the austenite transforms to martensite on cooling to room temperature. This martensite is untempered and the toughness is much lower and the hardness much increased. It should also be mentioned that the 4%Ni 'per se' will increase the toughness when in solid solution.

Certain of these high Ni steels have been used for power plant in hydro-electric applications, but not at high temperatures. However, it has been suggested that austenite 'hardened' 12%Cr steels in which the austenite is pre-

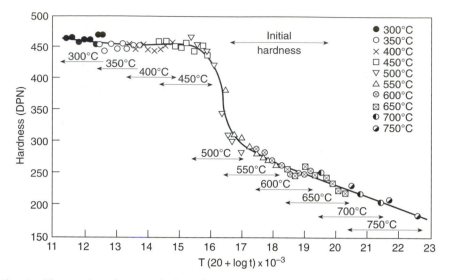

Fig. 4 Tempering characteristics of a 0.14%, 12Cr steel.

sent at the elevated operating temperature may be able to produce a power plant steel for conventional steam fired stations if the boundaries of the austenite particles are pinned by particles such as fine TiN. This would prevent the austenite particles coarsening and thus possible produce a highly stable microstructure subject to less degradation during operation. There are certain objections to such steels due to the inevitable carbides coarsening and leading to some degree of microstructural instability in the tempered martensite matrix.

THE TEMPERING CHARACTERISTICS OF 12%Cr STEELS

During tempering a 0.1%C, 12%Cr steel, retarded softening occurs up to 500°C followed by pronounced softening,[6] Fig. 4. At tempering temperatures above 550°C, the rate of softening decreases progressively. The microstructural changes can be correlated with these hardness effects. Up to about 350°C, tempering produces Fe_3C which grows from a fine dispersion to a 'dendritic' morphology and then to a plate-like Widmanstätten distribution, Fig. 5. During this process the Cr content of the Fe_3C increases to about 20%, at which there is some evidence, less than definite,[20] that Cr_7C_3 can form 'in situ' from the Cr-enriched M_3C, Fig. 6. Both these effects slow down the growth of the M_3C carbide and retard the softening. Also, at about 450°C tempering produces fine needles of M_2X which is predominantly $Cr_2(CN)$ in unalloyed 12%Cr steels, Fig. 7. This further helps to retard softening, but is not sufficiently intense to cause secondary hardening, although the M_2X forms by separate nucleation in the matrix. Above about 500/550°C the M_7C_3 and the M_2X phases coarsen appreciably, Fig. 8, and the hardness rapidly de-

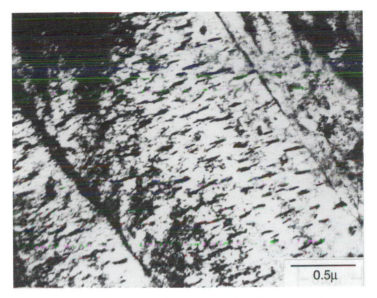

Fig. 5 Fe_3C precipitated in a 12%Cr type steel during tempering at 400°C.

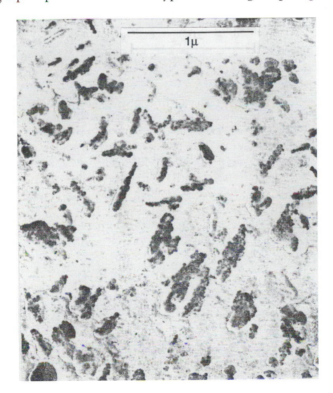

Fig. 6 Transformation of Fe_3C to Cr_7C_3 'in situ' during tempering a 12%Cr steel at 450°C.

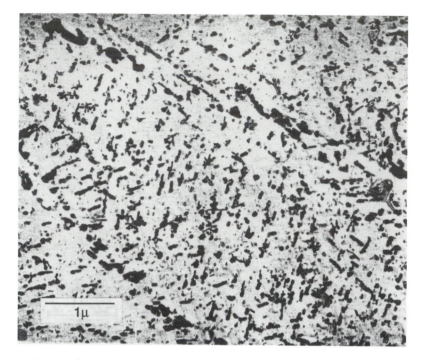

Fig. 7   Plates of Fe_3C transforming to Cr_7C_3 and fine needles of M_2X precipitated in the matrix of a 12%Cr steel tempered at 550°C.

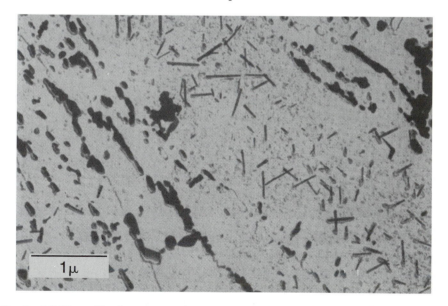

Fig. 8 M_2X needles in tempered martensite laths with Cr_7C_3 and $Cr_{23}C_6$ at the lath boundaries in a 12%Cr steel tempered at 700°C.

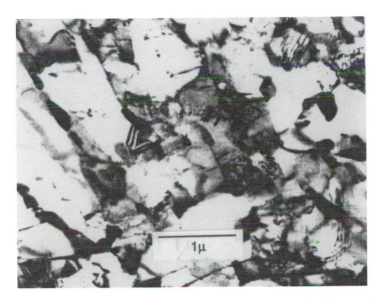

Fig. 9 $Cr_{23}C_6$ at tempered martensite lath boundaries and globular $Cr_{23}C_6$ at partially recrystallised ferrite boundaries in a 12%Cr steel heavily tempered at 750°C.

creases. At higher tempering temperatures the M_7C_3 and M_2X are replaced by Cr rich $M_{23}C_6$[5,6,21–23] which forms on the martensite plate and prior austenite grain boundaries, Fig. 9. The hardness decrease then becomes less rapid. Eventually as the $M_{23}C_6$ coarsens it loses its ability to pin the tempered martensite plate boundaries so that recrystallisation of the matrix ferrite slowly occurs, Fig. 10. When this takes place the hardness continues to decrease, but very slowly.

The M_2X phases is initially based on $Cr_2(CN)$, has a hexagonal structure and nucleates mainly on the many dislocations in the tempered martensite matrix. M_2X has the following orientation relationship with the matrix[5] which is precisely that for hexagonal Mo_2C phase (M_2X type) formed on tempering:

$$(0001)_{M_2X} || (011)\alpha \qquad (4)$$
$$[1120]_{M_2X} || [100]\alpha$$

It would appear that the presence of the 0.02/0.03%N, often occurring in 12%Cr steels, causes a hexagonal Cr_2N type phase (M_2X) to form preferably to the rhombohedral Cr_7C_3 phase, which itself is $Cr_{2.3}X$ and may be regarded as a defect lattice having interstitial vacancies. The N fills the interstitial vacant sites and changes the crystallography from rhombohedral (pseudo hexagonal) Cr_7C_3 to hexagonal $Cr_2(CN)$ which is the M_2X phase.

Fig. 10   Recrystallisation and growth of ferrite, with coarse $M_{23}C_6$ particles in a very heavily tempered 12%Cr type steel.

M_2X can dissolve many alloying elements such as Mo, W and V,[9,24,25] which increase its lattice parameter, and the associated coherency strains, and produce true secondary hardening with consequently increased tempering resistance, Figs 11, 12 and 13. Increased N will increase the volume fraction of M_2X and also produce secondary hardening and increased tempering resistance, Fig.14. In addition, an alloying element such as Si, dissolves in the matrix and lowers the matrix lattice parameter, increasing the coherency strains and increasing secondary hardening and tempering resistance.[1,5]

The M_2X phase, which in 12%Cr–Mo–V steel is $(CrMoV)_2(CN)$[24] is less stable than $M_{23}C_6$ at higher tempering temperatures, and gradually dissolves as $M_{23}C_6$ precipitates. The effect of alloying elements on the M_2X phase is the basis for the 12%Cr–Mo–V steels commonly in use. An interesting feature of the Ni invariably present is that it tends to accelerate carbide overaging and thus decreases the tempering resistance slightly. It is also interesting to note that the intensity of secondary hardening is increased by increasing C content, which increases the volume fraction of M_2X, Fig. 15. It is not usual, however, to use increased C contents which lower toughness and ductility.

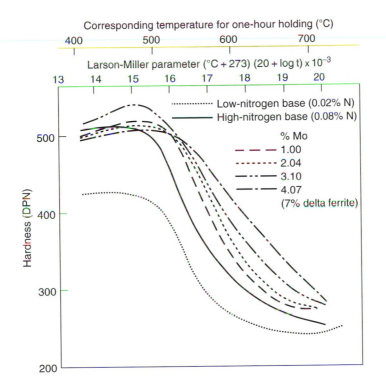

Fig. 11 Effect of Mo on the tempering characteristics of a 0.1%C, 12%Cr, 2%Ni steel.

Whilst V may dissolve in M_2X it also can precipitate as V(CN) at rather higher temperatures than does M_2X. It therefore increases the resistance to overageing Fig. 13, and therefore the tempering resistance. Nb behaves in a similar manner, precipitating Nb(CN) at even higher tempering temperatures, and so may Ta, producing Ta(CN) but also considerable solid solution strengthening. These *MX* phases precipitate by separate nucleation in the matrix, Fig16, and show a Baker-Nutting orientation relationship[26] with the matrix:

$$\{100\}_{MX} || \{100\}\alpha$$
$$<100>_{MX} || <110> \alpha \tag{5}$$

The MX phases are very stable, grow slowly, retard overageing and persist to the highest tempering temperatures.

The δ ferrite which may be present in incorrectly constitutionally balanced steel, also shows carbide precipitation. This δ ferrite frequently occurs in wrought products as polygonal ferrite grains, often connected into bands.

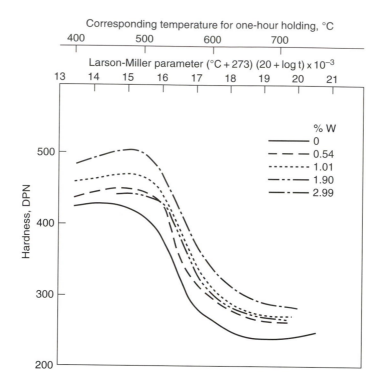

Fig. 12 Effect of W on the tempering characteristics of a 0.1%C, 12%Cr, 2%Ni steel.

Because it is rich in Cr, Mo and other ferrite forming elements, by virtue of partitioning effects, it often precipitates quite large needles of the M_2X phase,[27] which in this case is $(CrMo)_2(CN)$. These large needles of M_2X persist to high tempering temperatures.

Another feature of the presence of Nb, Ta and also Ti, is that they produce undissolved particles, due to the low carbide/nitride solubility, which pin the austenite grain boundaries and inhibit austenite grain growth. This improves the toughness by grain refinement, and by the same token improves the creep ductility. Due to the low solubilities of Nb(CN), Ta(CN), Ti C and TiN in austenite, it seems that too large quantities are added, e.g. 0.4%Nb, and that much of the Nb etc. are dissolved in the matrix thereby giving solid solution strengthening and retarding recrystallisation of the ferrite at the highest tempering temperatures.

It is possible therefore to increase the tempering resistance by additions of up to 2%Mo, or a small addition of W. Up to 1%Si may also be effective, and enhanced N may be used. All these affect the intensity of the M_2X secondary

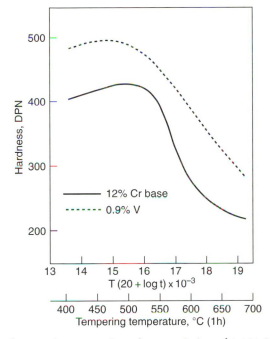

Fig. 13   Effect of V on the tempering characteristics of 0.1%C, 12%Cr steels.

hardening, and thereby the overaged hardness. If specific resistance to tempering at high tempering temperatures is desired, an addition of V or Nb (Ta) may be useful in producing fine precipitates of MX phases which form at much higher temperatures than does M_2X. It must be remembered, however, that the above elements are all ferrite formers and so the constitution will require careful balancing. Ni is the obvious balancing element, but too much should be avoided to prevent the Ac_1 being too greatly lowered. Co is a very useful balancing element, but is expensive. It does, however, have the advantage that it raises rather than depresses the M_s temperature.

PRECIPITATION HARDENING

Various precipitation hardening reactions can be used to increase the strength after tempering, although often with a decreased toughness and ductility. Frequently such reactions involve the precipitation of an intermetallic compound.[28,29] Copper is an example of age hardening, forming metallic copper precipitates, the maximum effect occurring at ~500°C, Fig. 17. Also up to 1.5%Al with 3%Ni can age harden by the precipitation of NiAl, Fig. 18, the maximum effect again occurring during tempering at ~500°C, Fig. 19. A similar effect can be achieved by Ni–Ti additions when the precipitating phase is NiTi. There are problems with such compositions because the maximum effect occurs at such low tempering temperatures that the toughness is not good, and

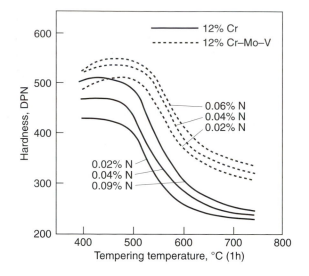

Fig. 14   Effect of N on the tempering characteristics of 0.1%C, 12%Cr and 0.1%C, 12%Cr–Mo–V steels.

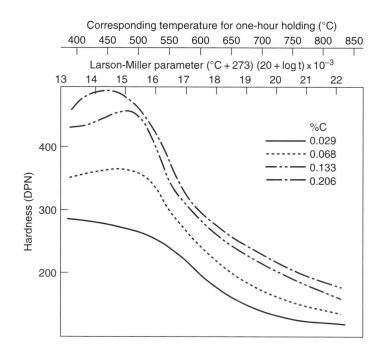

Fig. 15   Effect of C on the tempering characteristics of 12%Cr, 0.02%N steels.

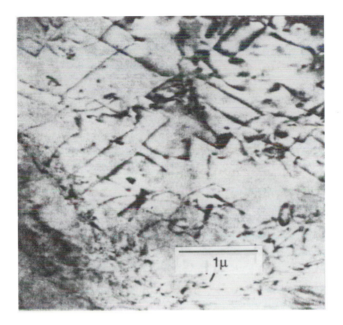

Fig. 16   Small MX precipitates formed in a 12%Cr–V–Nb steel precipitated on dislocations during high temperature tempering.

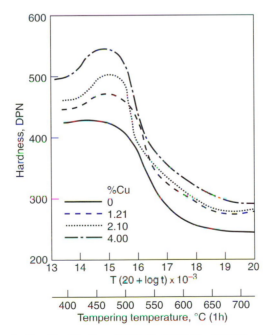

Fig. 17   Age hardening by Cu in a 0.1%C, 12%Cr, 2%Ni steel.

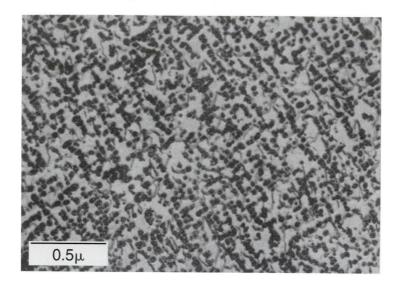

Fig. 18   NiAl precipitates in overaged 12%Cr steel containing 3%Ni, 1%Al.

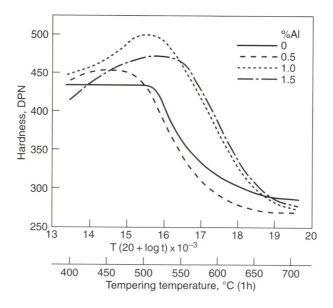

Fig. 19   Age hardening by NiAl in a 12%Cr steel containing 3%Ni, 1%Al.

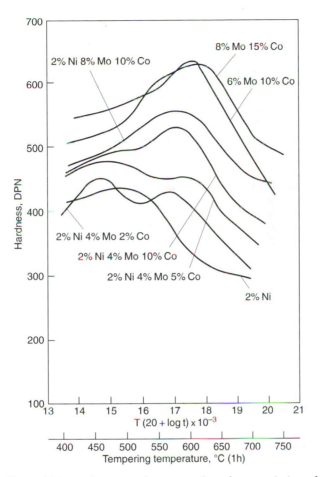

Fig. 20   Effect of Mo and Co on the tempering characteristics of 12% Cr steels showing the change in the ageing reaction with increasing Mo content.

higher tempering temperatures result in marked overageing and loss of strength. The steels however can be very much reduced in C which does not enter into the age hardening reaction, but some carbon is necessary to maintain a degree of strength in the overaged condition by carbide dispersion strengthening. The main difficulties are to maintain the constitutional balance and the correct transformation characteristics, which militate against the commercial application of such steels. The precipitation of M_2X can be intensified by up to 4% Mo or an equivalent amount of W. By increasing the Mo further and balancing the increased ferrite forming tendency by Co, which does not depress the M_s temperature, precipitation reactions can be produced which occur at higher tempering temperatures of 600/650°C. With increasing Mo and Co contents the whole level of the tempering curve is raised, Fig. 20, and the normal

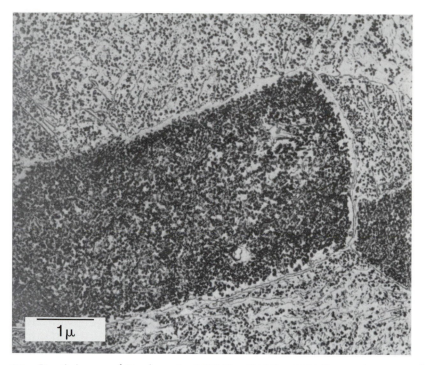

Fig. 21 Precipitates of R phase in 12%Cr, 6%Mo, 10%Co steel tempered at 700°C for 1 h. Note greater density of R phase particles in the delta ferrite due to Mo partitioning.

M_2X precipitation is replaced by that of an intermetallic compound based on the Mo–Cr–Co phase – R phase, Fig. 21[30] Moreover, the age hardening by R phase does not depend on C, so the steel can have a very low C content. The precipitation of the R phase which occurs with more than 4%Mo is intensified by the Co present. The lower C content of the steel means that the brittleness produced by the R phase is to some extent offset. Whilst R phase, a complex structure closely similar to σ phase, has been positively identified, as also has the Laves phase Fe_2Mo in the overaged condition, there have been less well substantiated reports of Fe_3Co and M_7Mo_6 phases. In fact there are possibly a variety of phases which exhibit certain similarities and can be formed with relatively small changes in steel composition. It has also been observed that similar precipitation effects can occur in steels containing W rather than Mo but a larger amount of W than Mo is required for an equivalent effect. In general the W steels seem to show rather better ductility and toughness than the Mo variant, possibly due to slower particle coarsening rates of the W intermetallic compounds compared with those of Mo. The effect of Mo content on the tempering conditions at which M_2X is replaced by the higher temperature R phase is shown in Fig. 22. Commercial steels, based particularly on the

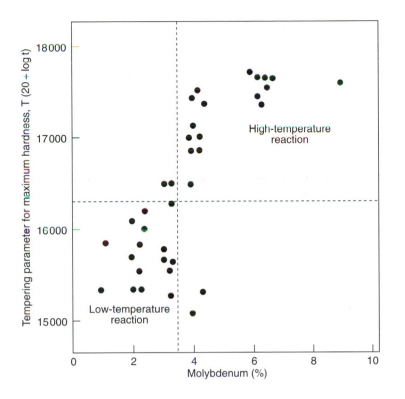

Fig. 22   Effect of Mo on the change from low to high temperature precipitation in 12% Cr steels.

Mo compositions, have been produced, but in general have found only slight industrial use.

The principles involved in both the intensification of M_2X precipitation and the use of intermetallic compound precipitation have led to some improved 12% Cr steels. Such steels contain various combinations of high N, Mo, W, V and Co and additions of Nb or Ta.[31-34] The problem with them seems to be associated with the precipitation of rapidly coarsening intermetallic compounds with overageing during prolonged testing or service. In particular the Laves phases $(FeCr)_2$ (MoW) seems to be detrimental to toughness, general ductility and high temperature creep ductility.

It is worth mentioning that the well known principle, so successfully exploited in simple secondary hardening steels and High Strength Low Alloy steels, in which the intensity of precipitation strengthening by carbides and nitrides of the V(CN) and Nb(CN) types is maximised by steels having the appropriate stoichiometric ratio, has not yet been applied to the 12% Cr steels. The reason for this may reside in the complex carbide precipitations which occur, and which render it extremely difficult to ascertain the effective carbon or nitrogen content to use for the calculations, particularly in the presence of undissolved stable carbides/nitrides.

MICROSTRUCTURAL STABILITY AND DEGRADATION

During tempering and especially during overageing at high tempering temperatures or long service times, various microstructural degradation effects can take place. Primary amongst these is the particle coarsening of the carbides. If a high tempering resistance is to be obtained, particularly at high tempering temperatures, and if the long term creep properties are not to deteriorate too markedly, the carbides produced during tempering should be slow to coarsen and stable so that they do not change and grow slowly during service. The phenomena which control particle coarsening are well understood[35] and described by the Wager equation:[36]

$$r_t^{\,3} - r_o^{\,3} = \frac{8\,\sigma\,D[M]V}{9RT} \cdot t \tag{6}$$

where r_o and r_t are the particle radii at time $t = 0$ and t respectively, σ is the surface energy of the particle-matrix interface, D is the diffusivity of the relevant atom species, $[M]$ is the concentration, i.e. the solubility, of the relevant atom species in the matrix in equilibrium with the particle, V is the particle molar volume, R is the gas constant and T is in K.

The most important factor is $[M]$, i.e. the solubility of the relevant atom species in the matrix in equilibrium with the particle, because this controls the flux of solute from the dissolving particles to the growing particles during particle coarsening. The lower this solubility, the more resistant will the particle be to coarsening. Thus stable, low solubility particles will resist coarsening. The stability of the various carbide phases increases as their enthalpy of formation increases, so even in the absence of well founded solubility data the enthalpy of formation can be used as an index of the stability and thus of the solubility. Some data for common carbides and nitrides is:[37]

	Phase	Approx. Enthalpy of Formation (KJmol⁻¹)
	Fe_3C	−10
	Cr_7C_3	−20
	$Cr_{23}C_6$	−25
	$Mo_2C\ (M_2X)$	−30
Decreasing	$W_2C\ (M_2X)$	−35
Solubility	$Cr_2N\ (M_2X)$	−40
	VC	−55
	NbC/TaC	−70

Phase	Approx. Enthalpy of Formation ($KJmol^{-1}$)
TiC	−95
NbN	−110
VN	−125
TiN	−170

These values are only approximate, especially as considerable solubility of elements in a given phase can occur and alter its stability and solubility. It can be seen, for example, that the M_2X phases have different stabilities depending on their composition, and there is extensive mutual solubility between them, as well as solubility for other elements, e.g. V.[24] Also a W based $M_{23}C_6$ will be less soluble than a Mo based $M_{23}C_6$ which in turn will be less soluble than $Cr_{23}C_6$.

Consequently, increased resistance to particle coarsening will be shown by carbides containing, in increasing order, Cr, Mo, W, V, Nb (Ta) and Ti. This explains much of the development of the 12%Cr steels because the above order more or less defines the chronological sequence in which alloying elements have been introduced into the steels. The search for microstructural stability seems therefore to have been logically followed, but care should be taken not to introduce too much of a particular alloying element as the excess over that required to form the particular carbide will simply dissolve in the matrix and could supply an increasing flux of atoms which would accelerate particle coarsening. Such excess solute would however tend to solid solution strengthen the matrix, but would not necessarily lead to microstructural stability. It will be noticed from the table of enthalpies of formation that the nitrides are more stable and less soluble than the corresponding carbide. Thus the use of enhanced nitrogen contents in 12%Cr steels is soundly based if microstructural stability and resistance to tempering is the objective.[31–34]

One of the features of the carbides, especially those formed at high tempering temperatures or long service times, is that they occur on the tempered martensite plate boundaries. They therefore pin the boundaries, preventing ferrite grain growth. The slower the particle coarsening rate the longer is the time before unpinning of the ferrite boundaries take place, with consequent secondary recrystallisation and ferrite grain growth. This pinning of the tempered martensite plate boundaries promotes microstructural stability and prevents the creep strength decreasing too catastrophically with extended service times. Also, because the grain boundary pinning prevents ferrite grain growth, the creep ductility and room temperature toughness does not deteriorate.

Athough the carbides formed at high tempering temperatures may be the most stable phases, there may be further carbide precipitation at lower service temperatures, and these carbides may not be the most stable of the phases. For example, at high tempering temperatures the solubility of the $M_{23}C_6$ may be such as to allow there to be enough C in solution for further carbide precipitation of the M_2X type to occur during service at lower temperatures. This has been observed[9,28] and can cause some increase in strength during the early stages of service with a loss of ductility and toughness.

In the currently popular highly alloyed 12%Cr steels containing appreciable amounts of Mo, W, Nb, V, Ta, Co, etc., there is always the tendency for precipitation of intermetallic compounds during service, even though the service temperature is below the tempering temperature at which the intermetallic compound does not precipitate due to the relatively short tempering time. Such phases are sigma, chi, Laves etc., and the precise intermetallic compound which forms will depend on the overall steel composition. In these steels it is not easy to predict the precipitation of a particular type of intermetallic compound by use of an electron vacancy number, as can be done in much more highly alloyed austenitic steels and Nimonic alloys.

EMBRITTLEMENT EFFECTS

There are many embrittlement effects which can occur, and which can only be discussed briefly. In essence they occur during prolonged service, but some can be the result of the initial tempering treatment. They fall into two types i.e. those giving rise to low creep ductility and those which lower the room temperature ductility and toughness and therefore may be important when a plant is taken out of service for maintenance. Some of the embrittlement effects manifest themselves in both types, i.e. high temperature and room temperature. An embrittlement which can occur on tempering is a form of temper embrittlement consequent upon specific residual impurities such as P and Sn which can produce both low creep ductility and a loss of room temperature toughness. Another form is that produced during tempering at 500/550°C in steels which show marked secondary hardening by M_2X and a consequent loss of room temperature toughness, Fig. 23. In steels which are not too resistant to tempering, at high tempering temperatures there may be secondary recrystallisation and grain growth of the tempered martensitic ferrite consequent upon carbide particle growth and loss of grain boundary pinning. This so-called 'upper' nose embrittlement gives low room temperature toughness but may also be reflected in decreased creep ductility. Another embrittlement effect can be caused by too coarse a prior austenite grain size which also results in low room temperature toughness and particularly a decreased creep ductility. Such a coarse austenite grain size may result from too high an austenitising temperature, possibly required to dissolve alloy carbides. It can be best obviated by pinning the austenite grain boundaries by fine particles of the TiN, Nb(CN), Ta(CN) types. It should be pointed out that in these transformable steels which are heavily dispersion strengthened the creep strength

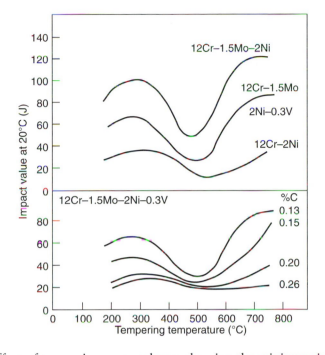

Fig. 23   Effect of tempering on toughness showing the minimum impact value at maximum secondary hardening.

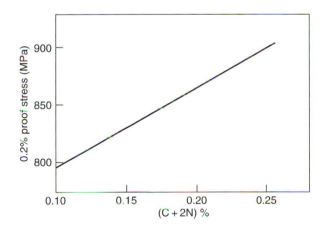

Fig. 24   Effect of C and N content on the 0.2% proof stress of 12%Cr steels after tempering 1h at 650°C.

is but little affected by a fine prior austenite grain size. By far the more insidious and less easily predictable embrittlements are those which occur during prolonged service. Mention has already been made of the additional fine carbide precipitation which can occur in service, and equally ferrite recrystallisation and grain growth may occur during service if the microstructural stability is poor. Both effects can cause a loss of room temperature toughness and in some cases lowered creep ductility. Especially in steels which contain some δ ferrite, 485°C embrittlement due to the precipitation of the Cr-rich α' phase may occur if cooling is slow through the embrittlement range. This may be exacerbated by the partitioning of ferrite forming elements to the δ ferrite because these elements and C and N increase the intensity of this form of embrittlement.[38] Also the precipitation of intermetallic compounds such as σ and χ and particularly the Laves phases, during prolonged service may decrease creep strength and ductility and also lower the room temperature toughness.[29] Considerable care is therefore needed to ensure the steels are not necessarily over alloyed. These intermetallic compounds often form preferentially in any δ ferrite, due to partitioning effects and further embrittle an already less than tough phase.

STRUCTURE-PROPERTY RELATIONSHIPS

Much work has been done on the mechanisms contributing to the strength of martensite but predictive equations have not been well developed, especially for the 12%Cr steels. Even less quantitative data is available on toughness and creep ductility.

The 12%Cr steels used for power plant are always tempered at high temperatures and there are some quantitative structure-property relationships for heavily tempered martensites which could be applicable. A distinction has to be made between structures in which the tempered carbides could occur only at the ferrite grain boundaries and those which also contain carbides within the tempered ferrite grains. The 12%Cr steels fall into the latter category and it has been shown[39] that the proof stress ($\sigma_{0.2}$) is controlled by the tempered ferrite grain size (d) through a Petch relationship and an additional strengthening (σ_p) due to precipitated carbides described by an Orowan type of equation:

$$\sigma_{0.2} = 77 + 23.9d^{-1/2} + \sigma_p \tag{6}$$

$$\sigma_p = \frac{0.015}{\lambda} \cdot \ln \frac{D}{2b} \tag{7}$$

where λ is the carbide spacing (μm), D is the carbide particle diameter (Å) and b is the Burgers vector of the slip dislocations (Å).

The constants in eqn (6) will require to be modified for use with 12%Cr steels and the validity of this type of approach still requires validation. A good quantitative relationship has been established between the 0.2%PS and the C and N contents of 12%Cr–Mo–V steels tempered at 650°C[8] Fig. 24.

$$0.2\%\,PS\,(MPa) = 710 + 770\,(C + 2N) \tag{8}$$

This equation simply reflects the increased strengthening due to the larger volume fraction of carbides/nitrides with increasing C and N contents.

In heavily tempered 12%Cr based martensitic steels it has also been shown[27] that an increase in the prior austenite grain size causes an increase in the ductile-brittle transition temperature of 23°C for a change in $d^{-1/2}$ of $-1mm^{-1/2}$. This is identical with results obtained by other workers on tempered martensite. Also it has been shown that the ductile-brittle transition temperature increases by 0.2/0.5°C for an increase in proof stress of 1Mpa[27] which again is identical with values obtained for other tempered martensitic structures. It is interesting to note that this effect is of the same order as the value of dispersion and dislocation strengthening on the ductile-brittle transition temperature of ferrite and would be expected as carbides coarsen and dislocations anneal out during tempering.

Despite the sparseness of quantitative data, it would seem that relationships based on Petch and Orowan equations may be applicable to describing the structure-property relationships in heavily tempered 12%Cr steels.

Even less quantitative data is available relating microstructure to creep strength and creep ductility and no predictive equations are available. It is possible to extrapolate creep strength and ductility to longer times by use of the very empirical extrapolation equations, but these give no indication of the effect of microstructure. It should be mentioned however that all the different microstructures will degrade during prolonged service exposure and will eventually reach the same low level of creep and rupture strength irrespective of the initial microstructure.

SUMMARY

This overview has attempted to describe and discuss the metallurgical phenomena associated with the evolution of the microstructure in alloyed 12%Cr steels used for high temperature power plant. Initially the compositional factors affecting the balance and control of the constitution have been described, together with the transformational aspects of the steels in terms of hardenability, martensite formation and re-austenitisation effects. The control of the transformation characteristics have been discussed.

The tempering characteristics of the basic 12%Cr steels have been considered together with the microstructural changes involved and the carbide precipitations and transformations which occur. The effect of alloying elements on these effects has been addressed from the viewpoint of increasing the tempering resistance and consequently the high temperature properties. Increasing the alloy content to produce yet further increases in tempering resistance and high temperature strength by the precipitation of various intermetallic compounds has been described.

The phenomena and mechanisms affecting microstructural stability and degradation during service are discussed in terms of carbide coarsening and the recrystallisation of the tempered martensite matrix and ferrite grain growth at high tempering temperatures or during prolonged service. Various embrittlement effects resulting from impurities, initial tempering treatment, partitioning to δ ferrite and especially the precipitation and growth of intermetallic compounds have been described. Finally a brief consideration of the structure-property relationships has been presented.

Throughout the review the metallurgical phenomena involved in the evolution of the microstructure have been related to the design of the steel compositions. Areas which have not yet been properly exploited or evaluated such as stoichiometric effects have been mentioned and the potential dangers of overalloying have been pointed out.

REFERENCES

1. J.Z. Briggs and T.D. Parker: *The Super 12%Cr Steels*, Climax Molybdenum Company, 1965, (Supplement 1983).
2. V.G. Rivlin and G.V. Raynor: *Int. Met. Review,* 1980, **248**(1), 21.
3. *Handbook of Stainless Steels*, D. Peckner and I.M. Bernstein eds, McGraw-Hill, New York, 1977.
4. K. Bunghardt, E. Kunze and E. Horne: *Arch. F.d. Eisenhuttenwesen*, 1958, **29**, 193.
5. F.B. Pickering: *Stainless Steels 84*, The Institute of Metals, 1985, 2.
6. K.J. Irvine, D.J. Crowe and F.B. Pickering: *J. Iron and Steel Inst.*, 1960, **195**, 386.
7. H. Schneider: *Foundry Trades J.*, 1960, **108**, 562.
8. F.B. Pickering: *Physical Metallurgy and the Design of Steels*, Applied Science Publishers, London, 1978.
9. F.B. Pickering: *Metallurgical Achievements*, Pergamon Press, Oxford, 1965, 109.
10. Y. Iwabuchi: *Trans. Jap Foundrymen's Soc.*, 1993, **12**, 69.
11. W. Steven and A.G. Haynes: *J. Iron and Steel Inst.*, 1956, **183**, 349.
12. G.H. Eichelman and F.C. Hull: *Trans. ASM*, 1953, **45**, 77.
13. P. Payson and C.H. Savage: *Trans. ADM*, 1944, **33**, 261.
14. F.B. Pickering: *Int. Met. Review*, Dec. 1976, **211**, 227.
15. S.R. Keown and F.B. Pickering: 'Niobium', *Proc. Of Int. Symp.*, H. Stuart ed., AIME, 1984, 1113.
16. K.J. Irvine: *J. Iron and Steel Inst.*, 1962, **200**, 820.

17. K.J. Irvine: *Journees Int. D. App. Cobalt*, 1964, 286.
18. C. Pickard and G. Nectoux: 'Stainless Steel Castings', *ASTM-STP 756*, 1982, 201.
19. W. Gysel, E. Gerber and A. Trautwein: *ibid*, 403.
20. F.B. Pickering: *Proc. Of 4ᵗʰ Int. Conf. on Electron Microscopy*, Berlin, Springer Verlag, Berlin, 1958, 665.
21. T. Fujita and T. Masumoto: *Tetsu to Hagane*, 1960, **46**, 1395.
22. B. Aronsson: *Steel Strengthening Mechanisms*, Climax Molybdenum Company, Michigan, 1969, 77.
23. F.B. Pickering: 'Precipitation Processes in Steels', *Special Report No. 64*, The Iron and Steel Institute, London, 1959, 23.
24. K.W. Andrews and H. Hughes: *J. Iron and Steel Inst.*, 1959, **194**, 304.
25. K.W. Andrews and H. Hughes: 'Precipitation Processes in Steels', *Special Report No. 64*, The Iron and Steel Institute, London, 1959, 57.
26. R.G. Baker and J. Nutting: *ibid.*, 1.
27. E.A. Little, D.R. Harries, F.B. Pickering and S.R. Keown: *Metals Tech.*, 1977, **4**(4), 208.
28. K.J. Irvine and F.B. Pickering: 'Metallurgical Developments in High Alloy Steels', *Special Report No. 86*, The Iron and Steel Institute, London, 1964, 34.
29. F.B. Pickering: 'Low Alloy Steels', *Publication No. 114*, The Iron and Steel Institute, London, 1968, 131.
30. D.J. Dyson and S.R. Keown: *Acta Met.*, 1969, **17**, 1095.
31. T. Fujita: *Metal Progress*, 1986, August, 33.
32. T. Fujita: *Trans. Jap. Inst. Met. (supplement)*, 1968, **9**, 167.
33. K. Oda and T. Fujita: *Tetsu to Hagane*, 1985, **71**(13), S1340
34. X. Liu, T. Fujita, A. Hizume and S. Kinoshita: *Trans. ISI Jap.*, 1986, **26**(3), B116.
35. F.B. Pickering: 'Encyclopaedia of Materials Science and Technology', *Vol. 7, Constitution and Properties of Steels*, F.B. Pickering ed., 1992, 362 et seq.
36. C. Wagner: *Z. Electrochem.*, 1961, **65**, 581.
37. R.W.K. Honeycombe: *Steels – Microstructure and Properties*, Edward Arnold, London, 1981.
38. M. Courtnall and F.B. Pickering: *Metal Science*, 1976, **3**(7), 273.
39. K. Onel and J. Nutting: *Metal Science*, 1979, 13(10), 573.

Precipitation Processes in Martensitic 12CrMoVNb Steels During High Temperature Creep

A. STRANG
GEC ALSTHOM Large Steam Turbines, Rugby, UK

V. VODÁREK
Vítkovice Research Institute, Ostrava, Czech Republic

ABSTRACT

Microstructural studies have been carried out on a series of 12CrMoVNb steels which exhibit sigmoidal creep rupture behaviour at test temperatures of 550°C and 600°C. The results indicate that this behaviour is accompanied by marked softening and microstructural changes in the material, due to precipitate coarsening and minor phase evolution in the alloy, leading to significant reductions in creep rupture strength. These processes result in the creep rupture properties of the material changing from being initially controlled by precipitation strengthening to being ultimately dependent on solid solution strengthening effects. The microstructural changes and precipitation sequences associated with the observed sigmoidal behaviour in these 12CrMoVNb steels are presented and discussed in this paper.

INTRODUCTION

Creep rupture studies on a series of tempered martensitic 12CrMoVNb steels, used for high temperature components in large steam turbine plant, have shown that these materials can be microstructurally unstable, particularly when used for extended periods of time at temperatures of 550°C or greater.[1] This behaviour manifests itself in the form of an inflexion in their creep rupture characteristics, in which a rapid reduction in rupture strength is accompanied by a corresponding increase in creep rupture ductility. This phenomenon is generally referred to as sigmoidal behaviour, and has been shown to be associated with progressive softening and precipitate coarsening effects in these materials, due to the combined effects of thermal exposure and plastic strain accumulation during the creep process.[2,3]

Although the phenomenon of sigmoidal behaviour is now well known, having been reported as early as 1963[4] by Bennewitz in more than 30 different steels with chromium contents ranging from 0 to 13%, the mechanisms and thermodynamics of the metallurgical processes responsible for the effect are not fully understood. Exposure temperature is known to be an important factor, with the inflexion in the stress rupture characteristics of the material occurring at progressively shorter durations as the test temperature is increased. Variations in composition are also known to have significant effects, with for example, creep rupture data on a wide range of creep resistant 12Cr steels indicating that increasing nickel contents also cause the sigmoidal inflexion point to occur at shorter test durations.[1,3]

Microstructural instability and degradation arising from factors such as these will not only affect high temperature material properties, but also the service lives of components manufactured from alloys prone to this type of behaviour. Furthermore, the effects of this type of material behaviour could potentially be wide reaching, particularly in circumstances where estimates of long term properties may have to be based on the extrapolation of relatively short term creep and creep rupture data. In consequence, a clear understanding of the processes leading to microstructural instability and degradation in these materials is of fundamental importance, especially to assist in the development of new metallurgically stable alloys suitable for applications in modern high efficiency steam plant, where service lives of 250 000 hours or greater at temperatures of 600°C or more are now routinely expected.

This paper describes the results of detailed metallographic studies which have been carried out on fractured creep rupture testpieces selected from a series of 12CrMoVNb steels known to exhibit sigmoidal creep rupture behaviour at 550°C and 600°C and discusses this behaviour in terms of the microstructural changes observed in these materials and their likely effect on long term high temperature material properties.

MATERIALS AND EXPERIMENTAL PROCEDURES

Materials
The two casts of 12CrMoVNb steel studied in this investigation were selected from a series of different casts of this alloy, all of which exhibited sigmoidal creep rupture behaviour when tested to durations of up to 100 000 hours at 600°C, Fig. 1. The selected casts exhibited the highest and lowest creep rupture strengths at these temperatures, as well as having the highest and lowest nickel contents of the series. The highest strength material (Cast A) had been solution treated at 1150°C and AC, followed by tempering for 6 hours at 650°C AC, while the lower strength material (Cast B) had been solution treated at 1165°C and AC, followed by tempering for 4 hours at 675°C AC. Details of the chemical compositions of these materials are shown in Table 1.

In the 'as received' condition, Casts A and B both exhibited tempered martensitic microstructures with each material having a mean Vickers dia-

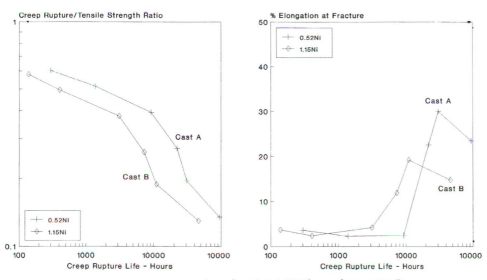

Fig. 1   Creep rupture properties of 12CrMoVNb steels at 660°C.

Table 1   Chemical Composition of 12CrMoVNb Steels (wt%)

Cast	C	Si	Mn	Ni	Cr	Mo	V	Nb	N_{tot}	Al_{tot}
A	0.16	0.28	0.74	0.52	11.20	0.61	0.28	0.29	0.074	0.006
B	0.14	0.13	0.88	1.15	11.74	0.50	0.29	0.30	0.064	0.003

mond hardness of 346 Hv_{10} and prior austenite grain sizes of 138µm and 28 µm respectively. No evidence of any δ-ferrite was found in either cast of material.

Experimental Procedures

Optical and analytical transmission electron microscopy studies were carried out on both casts of material in the 'as received' condition as well as on longitudinal and transverse microsections taken from each of the fractured creep rupture testpieces. In addition, detailed Vickers diamond hardness surveys were carried out at a load 10kg on longitudinal sections taken through the central axis of each of the fractured creep rupture testpieces, as well as on material in the 'as received' condition.

TEM and AEM studies were carried out on carbon extraction replicas and thin foils prepared from 'as received' and creep tested material, using a Philips CM20 transmission electron microscope fitted with an ultra thin window EDAX 9900 energy dispersive X-ray microanalyser. Quantitative analyses of minor phase particles extracted on replicas were carried out using PMTHIN software and the results normalised to 100% with respect to the elements detected. The carbon extraction replicas were prepared using the standard Smith and Nutting technique,[5] while the thin foils were prepared using a Struers Tenupol 3 twin jet electropolishing unit, operated at 60 volts with an electrolyte consisting of 5% $HClO_4$ in glacial acetic acid at room temperature.

RESULTS AND DISCUSSION

Creep Rupture Studies

The creep rupture properties of the series of 12CrMoVNb steels tested at 600°C are shown in Fig. 1. These indicated that sigmoidal creep rupture behaviour occurred in each of the casts tested, with inflexions being observed at test durations ranging from approximately 8 000 hours to 25 000 hours. In each instance the sigmoidal inflexion in the creep rupture strength characteristics was accompanied by a marked increase in rupture ductility. Optical microscopy revealed that the observed increase in creep rupture ductility was associated with a change in fracture mode from intergranular to transgranular fracture in the failed testpieces. Furthermore, as previously reported, there was a clear correlation between creep rupture strength and nickel content, with increasing nickel contents generally leading to reduced creep strength and increased creep rupture ductility at progressively shorter test durations.[1,3] On the basis of these data, the highest strength (Cast A) and lowest strength (Cast B) casts, respectively containing nickel contents of 0.52% and 1.15%, were selected for further detailed microstructural examination.

Hardness Studies

The results of the hardness studies conducted on both 'as received' and the heads and gauge length regions of the creep rupture tested materials are shown in Fig. 2. These results indicated that significant hardness reductions occurred in both casts of material during the creep process, with the degree of softening being greater in the creep strained gauge length portions of the testpieces compared with their thermally exposed unstrained heads. Furthermore, the rates of softening were observed to be noticeably greater in Cast B; the material with the higher nickel content.

These studies confirmed the results of previous investigations, namely, that the observed sigmoidal behaviour was associated with progressive marked softening of the materials with increasing exposure time at high temperature. These effects were particularly marked in the creep strained portions of the testpieces and in those casts with the higher nickel contents.[2,3]

Microstructural Studies

TEM studies of carbon extraction replicas and thin foils prepared from both 'as received' and creep rupture tested material from Casts A and B revealed

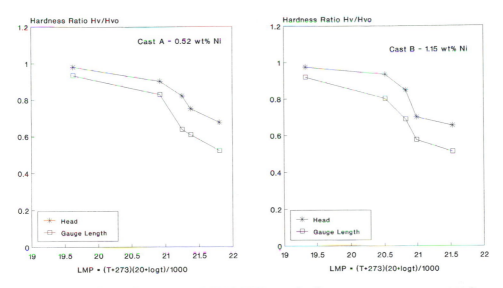

Fig. 2   Hardness changes in 12CrMoVNb steels after creep exposure at 600°C.

that large changes in the microstructure had occurred in both materials as a result of thermal exposure and creep strain effects during the creep process.

'As received' condition   In the 'as received' condition, extensive precipitation of coarse $M_{23}C_6$ particles was evident at the prior austenite and martensite lath boundaries in each cast of material. Large spherodised primary NbX particles were also present in each of the casts. Whilst these were mainly randomly distributed throughout the matrix, they also occasionally occurred in the form of stringers of smaller particles parallel to the original bar rolling direction. In each cast, a background of fine M_2X precipitate particles was also found to be present uniformly dispersed within the grains. However, much less of this precipitate was evident in Cast B, which contained more nickel and had been tempered at a higher temperature than Cast A. EDX analyses indicated that this fine intragranular precipitate was rich in chromium while electron diffraction studies established that it had a crystal lattice which was isomorphous with Cr_2N.[6] In Cast B, a small proportion of fine secondary MX particles was also found to be present in the matrix. These were shown by EDX microanalysis to be rich in vanadium and niobium suggesting the general formula (V,Nb)X. None of this phase was however found to be present in Cast A in the 'as received' condition. This was probably due to the lower temperature tempering treatment given to this material. Finally, thin foil electron microscopy studies of each material indicated that in the 'as re-

Table 2 Identification of Minor Phases Present in Casts A and B in the 'As Received' Condition and after Creep Testing at 600°C

Material	Condition	Phases Detected					
	Exposure Period – hrs	NbX	$M_{23}C_6$	M_2X	(V,Nb)X	Z–Phase	Fe_2Mo
Cast A	As Rec	√	√	√	–	–	–
	300	√	√	√	–	–	–
	9,200	√	√	√	√	–	–
	22,050	√	√	(√)	√	√	–
	30,750	√	√	–	–	√	–
	94,000	√	√	–	–	√	√
Cast B	As Rec	√	√	√	(√)	–	–
	140	√	√	√	√	–	–
	3,110	√	√	√	√	–	–
	7,245	√	√	–	–	√	–
	11,080	√	√	–	–	√	–
	46,155	√	√	–	–	√	–
	74,795	√	√	–	–	√	–

ceived' condition the dislocation densities were very high with no evidence of subgrains being found to be present in either cast of material. The results of the identification of the minor phases found in Casts A and B in the 'as received' condition, are summarised in Table 2, while typical electron micrographs of their microstructures are shown in Figs 3a–d.

Creep tested condition Electron microscopy studies of the heads and creep strained gauge length regions of the fractured creep rupture testpieces indicated that significant changes in microstructure had occurred in both casts with increasing exposure time at temperature and under stress. This was particularly noticeable in the case of the $M_{23}C_6$ particles located at the prior austenite and martensite lath boundaries and the fine M_2X precipitated in the matrix. Although no differences were found in the precipitation sequences of the minor phases between the heads and the creep strained gauge lengths of the testpieces, the $M_{23}C_6$ particle coarsening and M_2X dissolution rates were consistently found to be greater in the creep strained areas. Furthermore, in the materials investigated, increased $M_{23}C_6$ particle growth rates were also found with increasing test temperatures and nickel contents. Typical electron micrographs illustrating these changes in Cast A are shown in Figs 4a–d.

The metallographic studies also indicated that in both casts of material the

a) Cast A – 'as received'

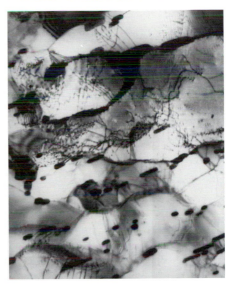

b) Cast A – after creep exposure
for 30 750 hours

|__ 0.5 μm __|

c) Cast B – 'as received'

d) Cast B – after creep exposure
for 46 155 hours

|__ 0.5 μm __|

Fig. 3a–d Microstructure of Casts A and B of 12CrMoVNb steel in the 'as received' condition and after creep exposure at 600°C.

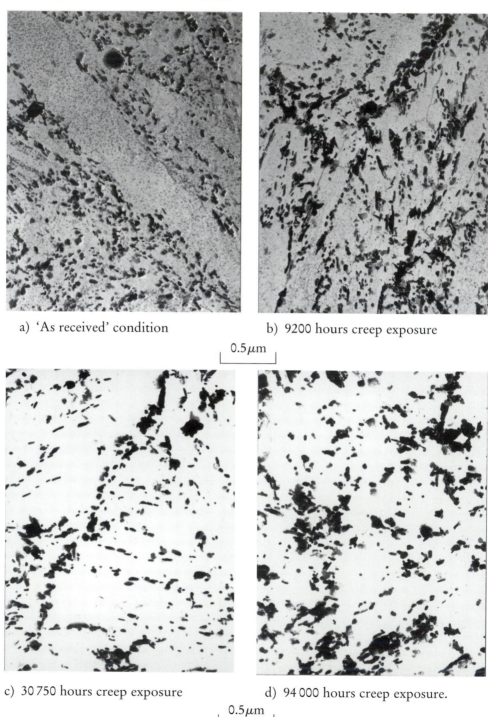

a) 'As received' condition

b) 9200 hours creep exposure

0.5 μm

c) 30 750 hours creep exposure

d) 94 000 hours creep exposure.

0.5 μm

Fig. 4a–d Microstructural changes in Cast A after creep exposure at 600°C.

fine M_2X precipitate was thermodynamically unstable. Whilst progressive dissolution of this phase occurred in both casts, particularly with increasing exposure time at 600°C, the rate was found to be significantly faster in Cast B, which contained the higher nickel content. In both casts dissolution of the

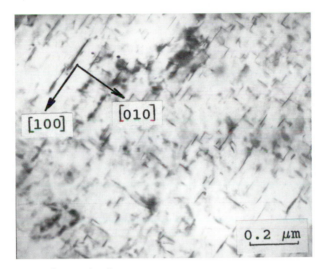

a) Secondary (V,Nb) X platelets

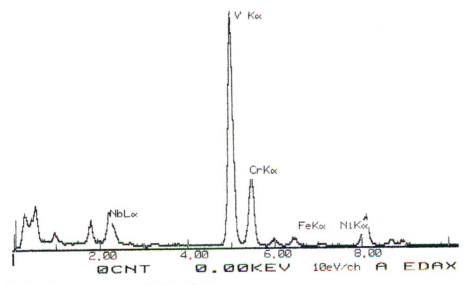

b) EDX spectrum of (V,Nb)X phase

Fig. 5a–b   Secondary (V,Nb)X phase precipitation in 12CrMoVNb steel after creep exposure for 9200 hours at 600°C.

fine M_2X was accompanied by the precipitation of fine secondary MX particles which were rich in vanadium and niobium and had the general formula $(V,Nb)X$. Furthermore, this phase had precipitated in the form of platelets in accordance with the well known Baker-Nutting orientation relationship, Figs 5a and b,[7] viz.,

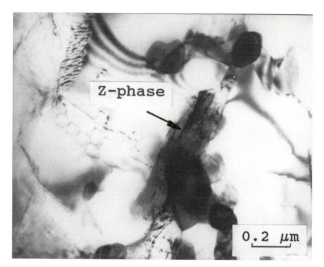

a) Z-phase particles

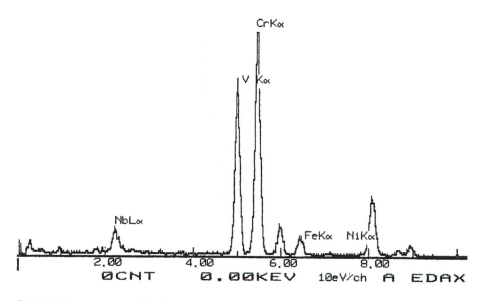

b) EDX spectrum of Z-phase

Fig. 6a–c Z-phase precipitation in 12CrMoVNb steel after 30 750 hours creep exposure at 600°C.

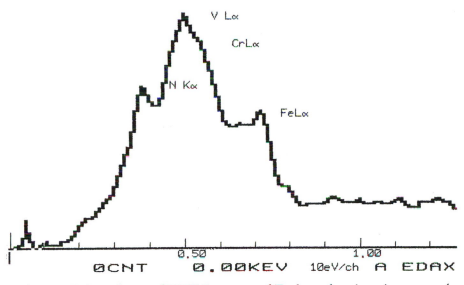

Fig. 6c   Enlarged part of EDX Spectrum of Z-phase showing nitrogen peak

$$(100)_{MX} // (100)_{\alpha}$$
$$[011]_{MX} // [010]_{\alpha}$$

The secondary MX precipitate was also found to be a transient phase, which redissolved following further creep exposure at 600°C, with the rates of dissolution once more being greater in the high nickel cast of material.

In both casts of material, the fine MX and M_2X precipitates were gradually replaced by a new plate-like phase which was shown by EDX analysis to be rich in vanadium, niobium and chromium, Figs 6a–c. Examination of the lower energy part of the EDX spectrum for this phase also revealed the presence of nitrogen thus suggesting that it was a complex nitride, Fig 6c. Convergent beam electron diffraction studies revealed the presence of two and four fold symmetry axes in the phase, while further selected area electron diffraction analysis indicated that the phase had a tetragonal unit cell with lattice parameters, a_o = 0.286 nm and c_o = 0.739 nm, Figs 7a–b. Furthermore, analysis of [001] orientation electron diffraction patterns revealed the absence of {hko} reflections consistent with the condition, h + k = 2n + 1, where n is an integer. This suggests that the space group P4/nmm is appropriate for this phase.[8] These results indicate that this phase is isomorphous with Z-phase; a complex nitride with ideal stoichiometry NbCrN which has previously only been observed in CrNi austenitic steels containing niobium and nitrogen.[9,10] The principal difference between this new phase and Z-phase is considered to be the substitution of niobium by vanadium resulting in a composition of the form Cr(V,Nb)N. This substitution qualitatively explains the observed dif-

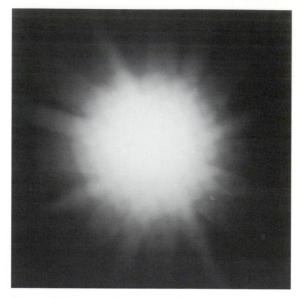

a) Convergent Beam ED pattern

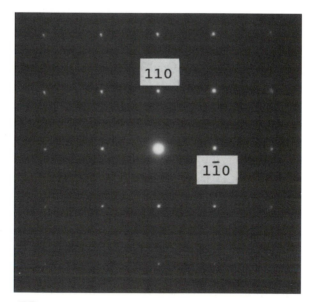

b) Selected Area ED pattern

Fig 7a–b   [001] electron diffraction patterns of Z-phase.

Table 3   Comparison of Z-Phase, $M_{23}C_6$ and NbX Interplanar Spacings

Phase	Interplanar Spacings (Å) and Reflections					
Z–Phase	2.66 (101)	2.26 (102)	2.02 (110)	1.85 (103)	1.77 (112)	1.56 (113)
$M_{23}C_6$	2.655 (004)		2.044 (333), (115)	1.878 (044)	1.770 (006), (244)	
NbX						1.580 (022)

ferences in unit cell dimensions on the basis of the relative atomic radii of the vanadium and niobium atoms, viz., $r_V = 0.135$ nm and $r_{Nb} = 0.146$ nm.[11]

The interplanar spacings of the modified Z-phase over the range of 2.66 nm to 1.56 nm are shown in Table 3, where they are compared with the d-spacings for $M_{23}C_6$ and NbX. Within this range of d-spacings it is clear that, with the exception of the 2.26 nm reflection, all of the other diffraction lines for the modified Z-phase are very close to the interplanar spacings for either $M_{23}C_6$ or NbX. Examination of X-ray spectra, previously obtained on extracts of minor phases prepared from 'as received' material and fractured creep rupture testpieces from Cast A tested at 550°C and 600°C, revealed diffraction lines indicating the presence of $M_{23}C_6$ and NbX phases in this material.[12] However, contrary to the metallographic findings, no evidence of M_2X was found in any of the X-ray spectra from these extracts. This was probably due to the amounts of this phase in the electrolytic extracts being too small for detection by this technique. However, an additional single strong asymmetric X-ray diffraction peak corresponding to an interplanar spacing of d = 2.26 nm was found to be present in the minor phases extracted from the long term creep rupture tests conducted on Cast A at both 550°C and 600°C, Fig. 8a.[12] Whilst the presence of this additional strong diffraction peak indicated that a substantial amount of another minor phase had formed in this alloy during creep exposure at both 550°C and 600°C, identification of the phase was not possible on the basis of this single diffraction line. However, the 2.26 nm X-ray diffraction peak does correspond with the (102) reflection found for the modified Z-phase. Furthermore, as shown in Table 3, the remaining diffraction lines for this phase are very close to interplanar spacings for $M_{23}C_6$ and NbX. On this basis there is strong evidence to believe that the other diffraction lines associated with the 2.26 nm reflection in the X-ray spectrum are obscured by the diffraction lines for $M_{23}C_6$ and NbX. This explains why only one diffraction line for this phase was evident in the X-ray spectra for the minor phases extracted from this alloy. It is therefore believed that the diffraction peak occurring at 2.26 nm in the X-ray spectrum for Cast A is due to the presence of modified Z-phase and not to a Cr_2Nb Laves phase, as

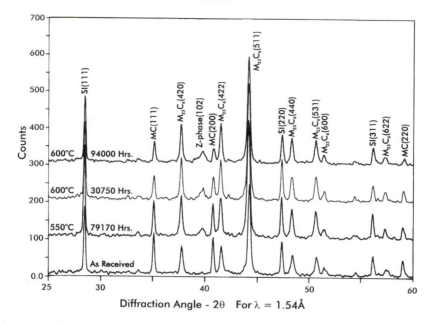

a) Spectra for Cast A

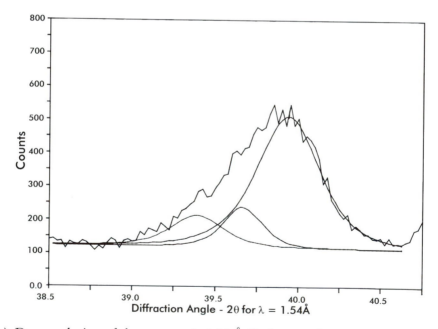

b) Deconvolution of the assymetric 2.26 Å Z–phase peak

Fig 8a–b   X-ray diffraction spectra for 12CrMoVNb steel in the 'as received' condition and after creep exposure at 550°C and 600°C.[12]

suggested in previous studies on this material.[12,13] Finally, while the electron microscopy studies have also revealed the presence of Fe_2Mo Laves phase in Cast A after creep exposure for 94 000 hours at 600°C, this phase also occurs in amounts which are too small to be detected by X-ray diffraction. Details of the minor phases detected in Casts A and B in the 'as received' and creep exposed conditions are summarised in Table 2.

The results of the quantitative EDX microanalyses carried out on the modified Z-phase found in Casts A and B are shown in Table 4. These analyses results indicate that although the Nb contents of each cast are similar, the average niobium content of the Z-phase is lower in Cast A than in Cast B. Furthermore, the high standard deviation values in these analyses results strongly suggest that the modified Z-phase is not chemically homogeneous. Chemical inhomogeniety would result in variations in the values of the lattice parameters for the modified Z-phase, and is probably the cause of the asymmetric form of the 2.26 nm diffraction peak observed in the X-ray spectrum for the minor phases electrolytically extracted from Cast A. Deconvolution studies of this peak suggest that it probably consists of a number of overlapping diffraction peaks thus inferring that several compositional variants of the modified Z-phase exist, Fig. 8b.[14] This is supported by the differences in the compositions of this phase between Casts A and B shown in Table 4 and raises questions regarding the 'equilibrium composition' for modified Z-phase in this alloy. Finally, the modified Z-phase was found to form at much shorter creep exposure durations in Cast B, which contains the higher nickel content, Table 2. This observation is consistent with the faster rate of dissolution of the M_2X and $(V,Nb)X$ phases also found in this cast of 12CrMoVNb material.

The results of the quantitative EDX microanalyses on the $M_{23}C_6$ particles in both steels in the 'as received' condition and after creep exposure at 600°C

Table 4 Chemical Compositions of Z-Phase in Casts A and B after Creep Exposure at 600°C

Material	Condition	Composition – normalised wt%					
	Exposure Period – hrs	V	Cr	Fe	Ni	Nb	Mo
Cast A	22,050	36.0 ± 4.6	47.0 ± 3.7	4.5 ± 0.6	-	11.1 ± 5.4	1.4 ± 0.5
	30,750	36.0 ± 3.3	45.9 ± 2.5	5.3 ± 0.7	-	11.3 ± 2.5	1.5 ± 0.5
	94,000	36.6 ± 2.3	44.9 ± 3.2	4.4 ± 0.6	-	12.9 ± 3.7	1.2 ± 0.2
Cast B	7,245	36.9 ± 4.2	41.4 ± 4.0	4.6 ± 0.7	-	15.9 ± 5.1	1.2 ± 0.2
	11,080	28.5 ± 1.8	43.4 ± 1.9	4.3 ± 0.7	0.3 ± 0.1	21.2± 2.3	2.3 ± 0.4
	46,155	29.3 ± 2.0	43.1 + 1.0	4.5 ± 0.6	0.4 ± 0.3	20.3 ± 3.1	2.4 ± 0.6
	74,795	22.2 ± 4.0	41.3 ± 1.9	2.8 ± 0.8	0.1 ± 0.1	32.0 ± 6.7	1.6 ± 0.4

Table 5   Chemical Compositions of the $M_{23}C_6$ Phase in Casts A and B in the 'As Received' Condition and after Creep Exposure at 600°C

Material	Condition	Composition – normalised wt%					
	Exposure Period – hrs	V	Cr	Fe	Ni	Nb	Mo
Cast A	As Rec	1.0 ± 0.2	67.8 ± 0.4	21.9 ± 0.7	0.7 ± 0.1	0.6 ± 0.1	8.0 ± 0.7
	300	1.7 ± 0.6	68.2 ± 0.4	20.9 ± 0.9	0.7 ± 0.1	0.5 ± 0.1	8.0 ± 0.8
	9,200	1.5 ± 0.3	69.7 ± 0.8	19.9 ± 0.7	0.7 ± 0.1	0.7 ± 0.2	7.6 ± 0.4
	22,050	1.1 ± 0.3	72.0 ± 0.4	16.4 ± 0.6	0.9 ± 0.2	0.6 ± 0.4	9.0 ± 0.7
	30,750	0.6 ± 0.1	73.5 ± 1.0	14.5 ± 1.2	1.0 ± 0.1	0.3± 0.1	10.0 ± 0.5
	94,000	0.6 ± 0.1	74.7 ± 0.7	14.9 ± 0.6	0.8 ± 0.1	0.5 ± 0.2	8.6 ± 0.3
Cast B	As Rec	1.1 ± 0.1	65.6 ± 1.1	24.1 ± 1.2	1.1 ± 0.2	0.6 ± 0.1	7.5 ± 0.8
	140	1.2 ± 0.2	67.3 ± 0.9	22.5 ± 1.9	1.0 ± 0.2	1.0 ± 0.5	7.1 ± 0.6
	3,115	1.0 ± 0.1	69.7 ± 0.4	20.9 ± 0.5	1.0 ± 0.1	0.7 ± 0.1	6.7 ± 0.2
	7,245	1.1 ± 0.3	69.8 ± 1.0	20.4 ± 1.3	1.0 ± 0.1	0.4 ± 0.2	7.3 ± 0.7
	11,080	0.8 ± 0.1	70.6 ± 0.7	20.0± 0.6	1.2 ± 0.2	0.5± 0.1	6.9 ± 0.8
	46,155	0.7 ± 0.1	73.3 ± 0.6	15.9 ± 0.9	1.4 ± 0.1	0.5 ± 0.1	8.2 ± 1.4
	74,795	0.9 ± 0.2	73.3 ± 0.3	16.1 ± 0.9	1.0 ± 0.3	0.7 ± 0.3	8.0 ± 0.9

are summarised in Table 5. These results confirm that the trends in the compositional changes observed in this phase are in accordance with the THERMOCALC predictions for this alloy and previous studies conducted on other creep exposed 12CrMoV steels.[15,16] Whilst the amount of nickel present in the steel appears to have had a significant effect on the coarsening rate of the $M_{23}C_6$ precipitate, the EDX microanalyses results indicate that only a negligible amount of this element is present in this phase. In both casts studied the nickel content of the $M_{23}C_6$ phase was similar to that contained in the basic steel and in each case, the amount present in the phase was found to be virtually independent of the duration of creep exposure. Finally, whilst no evidence of any M_6X precipitation was found in either cast of steel after creep testing at 600°C, other studies have shown that this phase is present in the high nickel heat after approximately 20 000 hours creep exposure at 550°C. Furthermore, EDX analyses indicated the presence of large amounts of nickel in this phase, an observation which is in good agreement with the results of other studies on 12CrMoV steels in which M_6X was found after long term creep exposure at 550°C.[17]

DISCUSSION
The detailed metallographic examination of the two casts of 12CrMoVNb

steel studied in this work has shown that in the 'as received' condition their microstructures consist of tempered martensite containing $M_{23}C_6$ carbides precipitated along both prior austenite and martensite lath boundaries with fine M_2X precipitates dispersed throughout the matrix together with large randomly dispersed spherodised particles of primary NbX.

During the course of creep exposure, particularly at 600°C, the intragranular M_2X precipitates progressively dissolved. This process was accompanied by the simultaneous precipitation of plate-like MX particles which were shown by EDX microanalysis to be rich in vanadium and niobium. This proved to be a transient phase which redissolved during further creep exposure at 600°C. Quantitative X-ray diffraction studies on the minor phases electrolytically extracted from Cast A demonstrated that the primary NbX particles were also being progressively dissolved during creep exposure, with the overall content of this phase being reduced to approximately 50% of that measured in the 'as received' condition after 94 000 hours creep exposure at 600°C.[12] All of these phases were gradually replaced by a more stable nitride phase which was rich in chromium, vanadium and niobium. This has been shown to be a modified form of Z-phase (CrNbN), which is an equilibrium nitride phase found in CrNi austenitic steels containing niobium and nitrogen. In the case of the 12CrMoVNb steels, the Z-phase formed as large plate-like particles in the matrix. The easy growth of Z-phase particles in this form in the ferrite matrix of this steel is probably due to the small degree of misfit between the ferrite lattice and the basal plane of the modified Z-phase. This is a reasonable assumption given that $a_{0Z-phase} = 0.286$ nm and for an α-ferrite matrix containing approximately 9 At% Cr , $a_{0FeCr} = 0.2864$ nm.[18]

The results of the metallographic studies on the high and low nickel 12CrMoVNb steels creep tested at 600°C indicate that the thermodynamically stable carbide phase present is $M_{23}C_6$. Two further equilibrium phases have also been observed to form in these alloys at 600°C, namely Fe_2Mo Laves phase and the complex nitride Cr(V,Nb)N, viz., the modified form of Z-phase. Due to the low molybdenum content of this steel, ~0.6 wt%, very little Laves phase forms and consequently, the presence of this phase has a negligible effect on the mechanical properties of the material. Apart from $M_{23}C_6$, modified Z-phase is therefore by far, the most important of the other equilibrium phases present. Since this is also a coarse phase, its presence has very little effect on the creep properties of the material in terms of precipitation strengthening. However, the processes associated with the formation of Z-phase have a significant effect on the evolution of the high temperature mechanical properties of the material. As Z-phase is precipitated, the other finely dispersed nitrides and carbo-nitrides which initially precipitation harden the material simultaneously dissolve. Indeed, the driving force for the dissolution of these finely dispersed phases is considered to be the precipitation of the modified Z-phase,

Table 6 Precipitation Sequences in 12CrMoVNb Alloys During Creep Exposure at 600°C

Material	Condition	
	As Received	Creep Exposed at 600°C
Cast A 0.52 wt% Ni	Primary Nbx $M_{23}C_6$ M_2X Secondary MX Z-Phase Laves Phase – Fe_2Mo	
Cast B 1.15 wt% Ni	Primary Nbx $M_{23}C_6$ M_2X Secondary MX Z-Phase Laves Phase – Fe_2Mo	

10   10^2 10^3 10^4 10^5

Creep Exposure Duration – Hours

which in turn is held to be the equilibrium nitride phase at 600°C in this particular martensitic creep resistant 12Cr alloy. The precipitation sequences associated with this process are summarised in Table 6.

M_2X and secondary MX phases occur as fine particles uniformly distributed throughout the matrix and initially contribute to the creep strength of the material by acting as barriers to dislocation movement at high temperatures, i.e., classical precipitation strengthening. Modified Z-phase occurs as large plates within the matrix and does not contribute to creep strengthening by this process. Initially, the $M_{23}C_6$ precipitate also contributes to the creep strength of the alloy by precipitation strengthening. However, as the $M_{23}C_6$ precipitate coarsens during creep exposure, unpinning of the prior austenite and martensite lath boundaries progressively occurs thus allowing recovery of the microstructure to take place. Evidence of this process was clear in terms of the much reduced dislocation densities and extensive subgrain networks found in the creep exposed materials compared with these materials in the 'as received' condition, Figs 3c and d. The combination of $M_{23}C_6$ coarsening, dissolution of fine M_2X and MX precipitates, and formation of Z-phase, lead to the observed reductions in creep rupture strength, and are directly responsible for the sigmoidal creep rupture behaviour of this type of steel. Thus the creep strength, which is initially dependent on precipitation hardening in these alloys, is progressively reduced due to a combination of particle coarsening, matrix softening and general microstructural degradation and, after the sigmoidal inflexion, becomes totally dependent on solid solution hardening effects in the material. These processes also account for the change from intergranular to transgranular fracture observed in the creep rupture testpieces after the sigmoidal inflexion.

Although the same precipitation sequences were observed in both of the steels investigated, the kinetics of these changes were found to be much faster in Cast B, which contained the higher nickel content. This was clearly demonstrated in the results of the creep rupture tests conducted at 600°C by the occurrence of the sigmoidal inflexion and increased creep rupture ductilities at approximately 8 000 hours duration compared with 25 000 hours in the case of Cast A with the lower nickel content. Also, as is evident from the results shown in Tables 2 and 6, the dissolution of the M_2X and MX phases and the precipitation of Z-phase occurred at significantly shorter creep exposure times in the high nickel cast. This provides further evidence that there is a clear correlation between the rate of microstructural degradation and nickel content in this type of steel and that increases in the nickel content of this alloy directly leads to an acceleration in the microstructural degradation process.

Finally, recent THERMOCALC studies on these 12CrMoVNb steels predict that the equilibrium phases expected to be present following long term exposure at 600°C, are MX and $M_{23}C_6$.[15] However, the metallographic results from this investigation indicate that the equilibrium phases actually present in these steels when exposed at this temperature are in fact primary MX, $M_{23}C_6$, Z–phase and Fe_2Mo Laves phase, with other evidence of M_6X also forming in the high nickel cast after long term exposure at 550°C.[17]

CONCLUSIONS

The results of microstructural studies conducted on two 12CrMoVNb martensitic steels with different nickel contents after normal quality heat treatment, i.e., the 'as received' condition, and subsequent long term creep exposure at 600°C can be summarised as follows, viz.,

1. In the 'as received' condition both casts of material exhibited tempered martensitic microstructures containing $M_{23}C_6$ precipitates at both prior austenite and martensite lath boundaries. In addition, a general dispersion of fine M_2X particles was present throughout the matrix of each cast with less of this phase being found in Cast B containing the higher nickel content. Primary NbX particles, which had not been dissolved during the quality solution heat treatment, were also found generally dispersed throughout the matrices of both casts of material.

2. During the course of subsequent long term creep exposure at 600°C, the M_2X precipitates were found to be thermodynamically unstable, gradually dissolving and being replaced by secondary MX particles, which were rich in vanadium and niobium. This precipitate was also found to be unstable at 600°C and dissolved with further creep exposure at this temperature. Both of these phases were replaced by large plate-like particles of a modi-

fied form of the complex nitride, Z-phase, which was shown by EDX microanalysis to be rich in chromium, vanadium, niobium and nitrogen. After 94 000 hours creep exposure at 600°C a small amount of Fe_2Mo Laves phase was also found to be present in this steel.

3. Whilst the modified Z-phase is considered to be the equilibrium nitride present in these steels after long term creep exposure at 600°C, this may not be true for steels exposed for long terms at 550°C, where, in the case of those containing high nickel contents, the presence of M_6X has been observed

4. Coarsening of the $M_{23}C_6$ carbides during creep exposure at 600°C leads to marked softening and recovery of the microstructure and this in conjunction with the formation of Z-phase and dissolution of the M_2X and secondary MX precipitates, leads to the observed sigmoidal behaviour and creep rupture strength reduction in this steel, with the process being accelerated in the steel with the higher nickel content.

5. In both of the steels studied the nickel content of the $M_{23}C_6$ particles was similar to that of the matrix and did not change with creep exposure at 600°C. However, coarsening of the $M_{23}C_6$ precipitate was observed to be more rapid in Cast B containing the higher nickel content.

6. The microstructural changes described in this paper occurred in both of the steels investigated. However the kinetics of the changes were affected by the composition and were significantly faster in the higher nickel cast.

REFERENCES

1. A. Wickens, A. Strang and G. Oakes: *Proc. Int. Conf. on Engineering Aspects of Creep*, Sheffield, 1980.
2. B.W. Roberts and A. Strang: *Proc. Conf. Refurbishment and Life Extension of Steam Plant*, Inst. Mech. Engs., London, October 1987.
3. T. Marrison and A. Hogg: *J. Metal Society*, 1972, **151.**
4. J.H. Bennewitz: *Proc. Joint Int. Creep Conf.*, Inst. Mech. Engs., New York/London, 1963.
5. E. Smith and J. Nutting: *British Journal of Applied Physics*, 1956, 7, 214.
6. S. Erikson: *Jenkontorets Ann.*, 1934, **118**, 530.
7. R.G. Baker and J. Nutting: 'Precipitation Processes in Steels', *ISI Special Report 64*, London, 1959.
8. *International Tables for X-Ray Crystallography*, The Kynoch Press, Birmingham, 1962.
9. H. Gerlach: *Proc. Conf. High Temperature Properties of Steels*, Iron and Steel Inst, Eastbourne, April 1966.
10. P.G. Stone: *Proc. Conf. High Temperature Properties of Steels*, Iron and Steel Inst., Eastbourne, April 1966, 505–511.

11. W. Rothery-Hume and G.V. Raynor: *The Structure of Metals and Alloys Fourth Edition*, Inst. of Metals, 1972.

12. H.K. Chickwanda: 'Microstructural Stability of 12CrMoVNb Power Plant Steels', *PhD Thesis*, Imperial College, London, January 1994.

13. H.K. Chickwanda, A. Strang and M. McLean: *Int. Conf. on Materials for Advanced Power Engineering*, Liege, October 1994.

14. R. Sweeny: *Private Communication*, Imperial College, London, 1996.

15. J. Hald: T U Denmark (ELSAM) Private Communication, 1995.

16. R.C. Thomson and H.K.D.H. Bhadeshia: *Met. Trans. A*, April 1992, **23A**.

17. A. Strang and V. Vodarek: *Unpublished Results*, 1996.

18. A.G.H. Anderson and E.R. Jette: *Trans. Amer. Soc. Met.* 24(375), 193.

The Development of 9% CrMo Steels from Steel 91 to E911

J. ORR AND L. WOOLLARD

British Steel plc, Swinden Technology Centre, Moorgate, Rotherham, S60 3AR

ABSTRACT

Steel 91, a 9% Cr 1% MoNbVN has become recognised as the base high strength ferritic/martensitic steel for elevated temperature service in place of some of the austenitic stainless steels, thus giving greater flexibility of operation for some critical boiler components. It has also found use in replacing some lower alloy steels which are unable to withstand effectively flexible operation conditions.

This paper reviews the metallurgy of Steel 91, by relating the microstructure to properties. The effects of heat treatment temperatures and composition changes which may be encountered in commercial and experimental materials are described emphasising the role of MX precipitates and solid solution strengthening.

The development, optimisation and stability of the now established E911 composition are discussed in terms of the microstructural parameters.

1. INTRODUCTION

Since 1936 the 9% Cr 1% Mo steel composition has been recognised as an important base for applications in technological industries such as oil production[1] and power generation. The first development beyond the basic composition was in Belgium in the late 1950s to produce a martensite + delta ferrite microstructure strengthened by niobium and vanadium.[2] This, however, became brittle due to Laves phase precipitation and therefore is not generally recognised today other than in France and Belgium. A later development, firstly in the USA and followed in Japan, was driven by a need for a high strength stainless steel for the US Fast Reactor but of lower cost and better stress corrosion resistance than Type 304. Out of an extensive research programme at the Oak Ridge National Laboratory (ORNL) grew the composition recognised today as Steel 91.[3] Similar alloy compositions were developed in Japan which were known as HCM9M, Tempaloy F9 and NSCR9 which were 9Cr 1–2MoNbV compositions.[4] None of the latter steels became recognised formally even in Japan, whilst Steel 91 has become known worldwide appearing in US, UK, German, French and draft European specifications or data sheets.[4]

The Steel 91 composition has itself become the basis for further steel developments in Europe and Japan, leading to steel grades such as E911, NF616, etc.

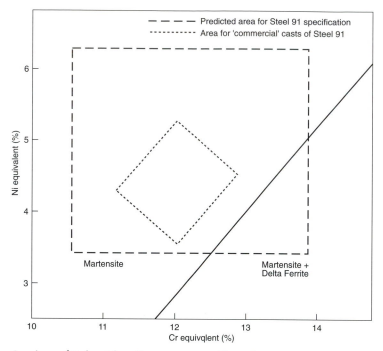

Fig. 1 Section of Schneider diagram to predict microstructures.

It is the purpose of this paper to define the basic metallurgical components of the Steel 91 type compositions and how these have been refined further to the 'advanced' 9–12% Cr steels, with specific reference to the steel E911.

2. THE METALLURGY OF 'STEEL 91'
The base composition of 9% Cr, 1% Mo means that steel products will be hardenable, i.e. fully martensitic, with dimensions up to the equivalent of 2000 mm diameter when air cooled. The relatively small additions of niobium and vanadium, as in Steel 91, do not cause any significant changes in the transformation from austenite to martensite.

2.1 Delta Ferrite
The specified ranges of composition for Steel 91, Table 1, indicate that delta ferrite can be present in the microstructure, but typical commercial compositions may be expected normally to give fully martensitic microstructures, Fig. 1. However, because of the closed γ-loop for high chromium steels it is also possible to form delta ferrite by raising the normalising/solution treatment temperature, as illustrated in Fig. 2.[5] This is similar to the situation found for basic 9% Cr, 1% Mo steels when subjected to simulated welding conditions, particularly when the silicon content is raised above ~0.5%.

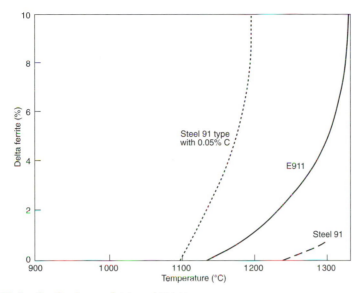

Fig. 2   Delta ferrite in steel 91 and E911.

Table 1   Typical steel compositions

Steel Name	Wt.%									
	C	Si	Mn	Cr	Mo	Ni	Nb	V	N	W
Steel 91	0.10	0.3	0.4	8.5	1.0	0.2	0.08	0.20	0.05	–
E911	0.11	0.2	0.4	9.0	1.0	0.2	0.08	0.20	0.07	1.0

2.2 Martensite Hardness and Tempering

The hardness of the martensite in Steel 91 after normalising from typically 1050–1100°C is found to be related to the sum of the carbon and nitrogen contents, rather than the former alone, reaching values of ~400 HV30.[6] During tempering there is a secondary hardening reaction, achieving a peak hardness of 300–350 HV30 after 1 hour at 650–690°C, Fig. 3. Although this has not been investigated directly, evidence from samples tempered at 750–780°C indicates that this is due to the precipitation of small particles of niobium/vanadium carbide/nitride. The hardness values after tempering treatments at 700–780°C are 270→200 HV30. The latter value represents the lower limit of hardness that can be achieved normally in Steel 91 since tempering at above about 800–825°C can result in exceeding the Ac_1 temperature and hence partial reversion to austenite and thence to untempered martensite on cooling to ambient temperature, Fig. 3. The actual Ac_1 temperature depends for example on the nickel content of the steel. For tube and pipe applications, which are welded and stress relieved before entering service, the highest possible Ac_1 is required so that initial tempering *and* stress relieving can be carried out usually within the range 750–780°C. Hence, the nickel con-

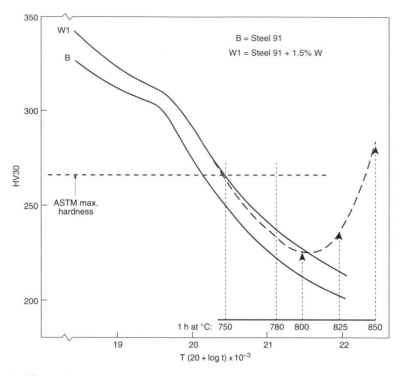

Fig. 3 Tempering curves.

tent is specified at ⩽0.4% and is usually below ~0.2%. For non-welded applications, e.g. rotors, lower tempering temperatures of typically 700–730°C,
are used to gain higher strengths and higher nickel contents are used to maintain a reasonable level of toughness.

The microstructures after normalising from 1050–1100°C and tempering at
750–780°C show prior austenite grain boundaries and sub-boundaries outlined by $M_{23}C_6$ particles, Fig. 4 and Table 2(a). Higher magnification examination shows that some of the $M_{23}C_6$ is present as small particles with sizes in
the range 0.2–0.5 μm, which may therefore contribute a dispersion strengthening component. Fine precipitates of NbC or V_4C_3/VN are also present
with sizes of typically 0.1 μm, Figs 5 and 6. The former are usually
rounded/sperical particles, Fig. 5, whereas the latter are as small plates or
rods, Fig. 6. Occasionally, particles with compositions approximately mid-
way between these two types, i.e. 40% V 40% Nb 14% Cr 6% Fe are observed, as either double or single particles, Fig. 7(a) and (b). However, after
testing at 600°C the definite formation of 'winged' particles reported by
others[7] is observed, Fig. 8.

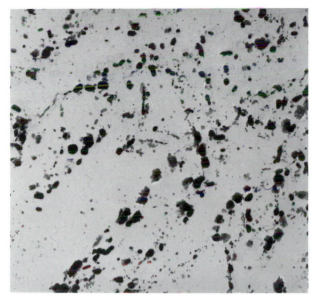

N1050 + T750

2 μm

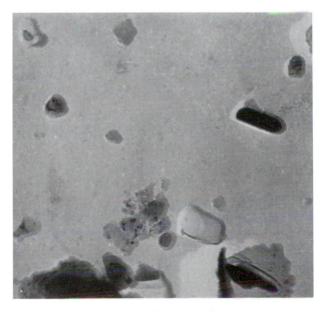

37 068 h − 650°C

0.2 μm

Fig. 4 $M_{23}C_6$ in Steel 91.

Table 2(a–e)   Compositions of metal fractions in precipitates in Steel 91

(a) $M_{23}C_6$

°C	%				
	Cr	Fe	Mo	W	
1050 + 750	68	26	6	–	B
	65	27	2	6	W
+ 600 (13 472–35 463 h)	69	24	7	–	B
	67	23	4	6	W
+ 650 (1429–15 616 h)	71	21	8	–	B
	67	19	4	11	W
+ 700 (3000 h)	68	16	8	8	W

B = Base Steel W = Tungsten Steel

(b)   **VN**

°C	%			
	V	Cr	Fe	Nb
1050 + 750	55	26	9	3
+ 600 (18 592 h)	68	26	3	3
+ 650 (2305–37 068 h)	68	18	8*	6*
+ 700 (3000 h)	80	12	2*	15*

* Not in all particles

(c)   **NbC**

°C	%			
	Nb	V	Cr	Fe
1050 + 750	90	8	1	1
+ 600 (18 592 h)	85	9	3	3
+ 650 (1429–19 244 h)	88	7	4	1
+ 700 (3000 h)	93	4	2	1

Table 2(a–e)   Compositions of metal fractions in precipitates in Steel 91
(Continued)

(d) Laves Phase

°C	%			
	W	Mo	Fe	Cr
600 (35 453 h)	55	12	21	11
650 (1429–11 075 h)	52	12	30	6
700 (3000 h)	47	13	33	6
NF616 10 000 h/650°C*	24	10	51	15

* See ref. 10

(e) VN/NbC Particles

°C	V	Nb	Cr	Fe
1050 + 750 + 600°C* + 700°C	40 27–40 21–58	39 41–57 29–39	14 12–16 13–27	7 3–6 0–1

* Generally found as 'winged' particles (see Fig. 8)

2.3 Prior Austenite Grain Size

The prior austenite grain size of martensitic steels, particularly those containing vanadium, is considered to be a reflection of their metallurgical stability.

The addition of niobium and vanadium to the base 9% Cr 1% Mo steel results in grain refinement for normalising treatments up to 1125°C, above which fairly rapid grain coarsening occurs, Fig. 9(a). Although no direct metallographic evidence is available, empirical studies indicate that it is the Nb(C,N) type particles which determine the grain size in Steel 91, Fig. 9(b).

3. EFFECT OF TIME AND TEMPERATURE ON PRECIPITATION IN STEEL 91 AND E911

3.1 Normalised and Tempered Condition

Steel 91 is the general name for the basic compositions 9% Cr 1%MoNbVN, Table 1, to which most of the foregoing descriptions refer. E911 is the composition developed from Steel 91, as a result of an ECSC sponsored research project,[6] in which the vanadium and nitrogen are controlled to closer limits than in Steel 91, and an addition of 1% W is made. In all other aspects the compositions of the two steels are similar, Table 1. In the normalised 1050°C + 750°C condition, the microstructure consists of tempered martensite.

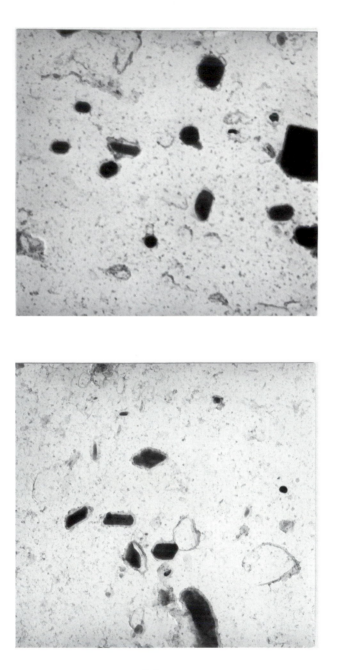

0.2 μm

Fig. 5 NbC PPTS.

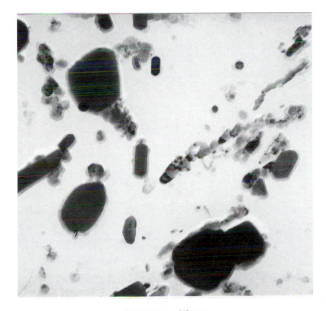

N1050 + T750

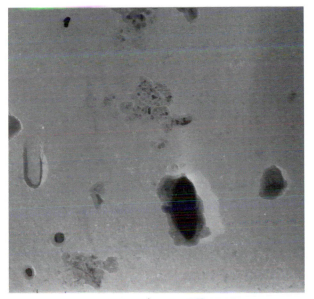

37 068 h − 650°C

0.2 μm

Fig. 6   VN PPTS.

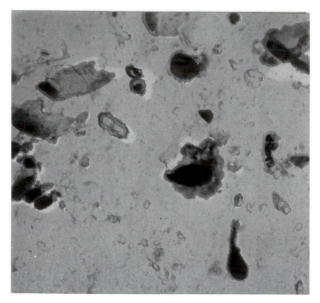

N1050 + T750

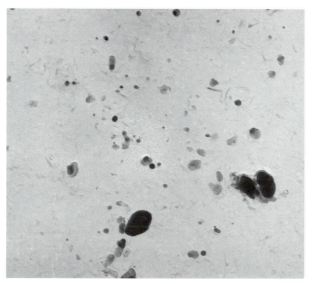

18 592 h – 600°C

⌐0.2 μm⌐

Fig. 7   (VNb) (CN).

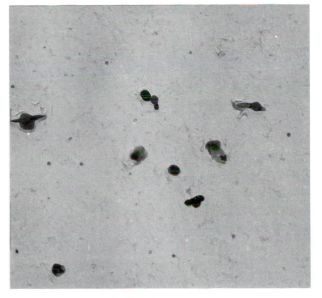

18 592 h – 600°C

0.2 μm

Fig. 8   (VNb) (CN) 'WINGS'

Steels with compositions relating to the base composition and with 1.5%W, which was the experimental forerunner to E911, all have similarly fine VN and NbC type particles in the normalised and tempered condition, Figs 5–7. The compositions of these particles are similar in both steel types, Table 2(b) and (c).

As indicated above, the contribution of strength from VN precipitation can be maximised, through stoichiometric balancing of the two major elements in this precipitate, as demonstrated in Fig. 10. The presence of a significant amount of chromium and some iron and niobium in the VN precipitates, Table 2(b), means that the stoichiometric balance may not occur at the theoretical ratio of 3.64, as indicated by the data in Fig. 10. By comparison the composition of the NbC particles are much closer to the 'pure' form, Table 2(c). It is considered, as indicated above, that it is the NbC particles which determine the grain size of Steel 91 and its derivatives, whilst it is VN that provides principally the precipitation strength. This conclusion is based on empirical evidence in which grain size is found to be related to variations in carbon in conjunction with niobium, see Fig. 9(b), whilst strength/hardness correlates with the vanadium to nitrogen ratio, Fig. 10, but not with niobium/carbon/nitrogen relationships. The generally recognised more refractory nature of NbC compared with VN in terms of solubility may also relate to the respective purity of these particles.

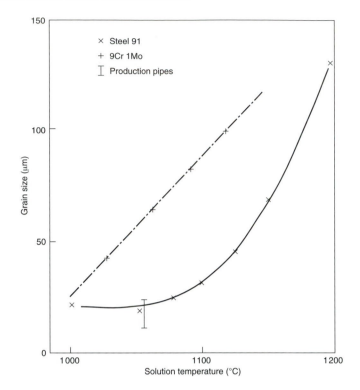

(a) Prior austenite grain sizes in Steel 91

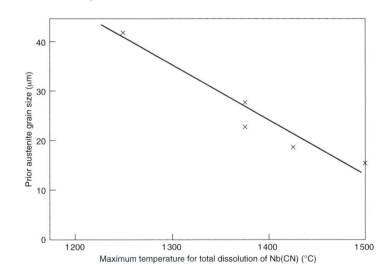

(b) Grain size/Nb(CN) solubility relationship

Fig. 9(a–b).

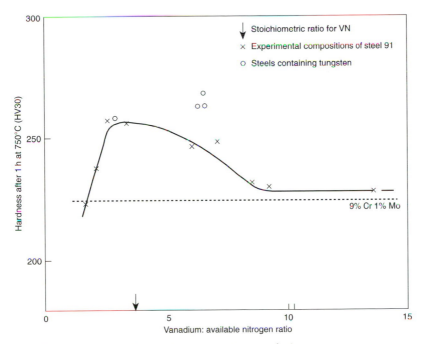

Fig. 10   Hardness vanadium: nitrogen ratio correlation

Since the NbC and VN particles are of similar size and sometimes found in conjunction with each other, Figs 5–8, it is considered that the former do also contribute some precipitation strength.

3.2 Test Duration Exposure
Samples from several specimens of Steel 91 and tungsten added steels, have been tested for up to 35 453 h at 600–700°C and subsequently examined by TEM to determine precipitate sizes, distributions and their metal fraction compositions. The results from these studies are included in Figs 5–8 and Table 2(a–d).

The results show that precipitates of $M_{23}C_6$, VN and NbC remain at least similar in size to those in the normalised and tempered conditions, Figs 5–8. However, some of the particles are even finer after exposure at 600/650°C, than in the normalised and tempered conditions thereby suggesting that further precipitation has occurred at these temperatures, but without significant coarsening of pre-existing particles. This effect results from the differential solubility at 750°C and 600–650°C of the constituent elements of these particles.

There is no significant change in composition of the NbC type particles, during exposure at 600–700°C, which are typically 90% Nb, 7% V, 2% Cr, 1% Fe, Table 2(c).

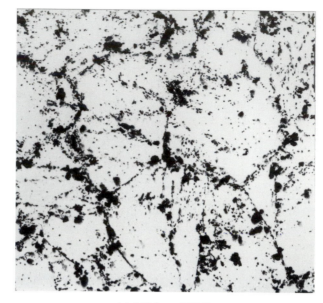

14 269 h – 650°C

5 μm

35 458 h – 600°C

0.5 μm

Fig. 11 Laves phase.

For the VN particles the composition changes from typically 58% V, 26% Cr, 9% Fe, 3% Nb in the normalised and tempered condition, with increases in vanadium and decreases in chromium, iron and niobium contents. After exposure at 650 and 700°C not all particles of VN contain iron and niobium, Table 2(b).

The $M_{23}C_6$ particles show only small changes in composition during exposure at 600–700°C, with decreases in iron and small increases in the chromium and molybdenum contents, Table 2(a).

In the tungsten added steel some tungsten is present in the $M_{23}C_6$ particles displacing an equivalent total amount of chromium + molybdenum in approximately equal proportions, Table 2(a).

Laves phases appears in the microstructure of the tungsten containing steel tested at 600, 650 and 700°C but is not present in the normalised and tempered condition. The particles of Laves phase are fairly large, appearing on the prior austenite grain boundaries, Fig. 11. The composition of this phase depends on the temperature at which it formed, Table 2(d). These data show a decrease in the tungsten (55→47%) and chromium (11→6%) contents with an increase in the iron content (21→33%) as the test temperature increased from 600 to 700°C, Table 3(d).

4. STRENGTHENING MECHANISMS IN STEEL 91 AND E911
The metallographic and empirical evidence indicates that in Steel 91 and in E911 steel, Table 1, there are several strengthening mechanisms which are summarised in Fig. 12.

4.1 Solid Solution Strengthening
The principal alloying elements in these steels are chromium, molybdenum and also tungsten, when added as in E911. Relatively small amounts of molybdenum and tungsten appear in the $M_{23}C_6$ and VN particles, Table 2(a) and (b). Therefore it can be assumed that all three elements provide mainly solid solution strengthening, though some dispersion strengthening by the smaller $M_{23}C_6$ particles may also contribute to the basic strength. This is indicated empirically by examination of the detail in Fig. 12 which indicates that the strength component up to the lowest contributed by the VN precipitation strengthening component is largely that due to the basic 9% Cr, 1% Mo composition. It is postulated also in Steel 91 that the fine NbC particles could also contribute some strength but this effect will be small relative to that of VN precipitation because of the low solubility of NbC at 1050°C, Fig. 13, which is the usual normalising temperature for Steel 91. There are data to show that by increasing the normalising temperature to 1100°C a 10% increase in creep strength of Steel 91 is gained.[6,8] This could be attributed to dissolution of more niobium, vanadium and the interstitial elements which precipitate on tempering.

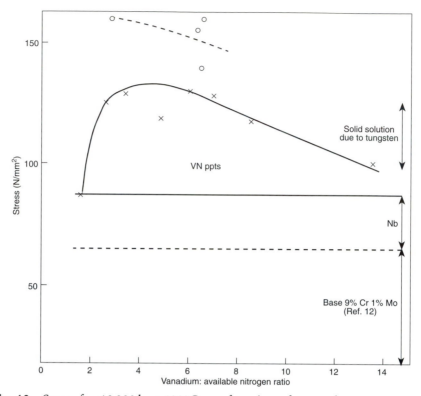

Fig. 12 Stress for 10 000 h at 600°C as a function of strength parameters.

The addition of 1–1.5% W also increases the creep strength by 10–15%, Fig. 12. A similar effect to that shown in Fig. 12 has been found for an increase in the chromium content over the range 8.2–10.5%, but is less stable at 600–650°C than that from tungsten.[8] The strengthening effect of tungsten is confirmed by others, albeit for only 1000–3000 h at 600°C, though the same author argues that this is not due to solid solution but rather to precipitation of grain boundary films of Laves phase and/or an increase in the lattice constant for VN by inclusion of some tungsten.[9] Other information suggests also that solid solution strengthening by tungsten is small.[10] There is no evidence from the present studies that precipitates other than $M_{23}C_6$ (and the latter forming Laves phase) contain tungsten, Table 2, nor of the thin films of Laves phase at prior austenite grain boundaries linking $M_{23}C_6$ particles as suggested by reference 9. An argument against the proposals for Laves strengthening is that the beneficial effect of adding tungsten is present in the normalised and tempered (1050 + 750°C) condition as indicated by the short duration creep tests, i.e. those less than ~3000 h as indicted by data for NF616 and Steel 91 for example. It is argued that Laves cannot form above ~720°C[10] and there-

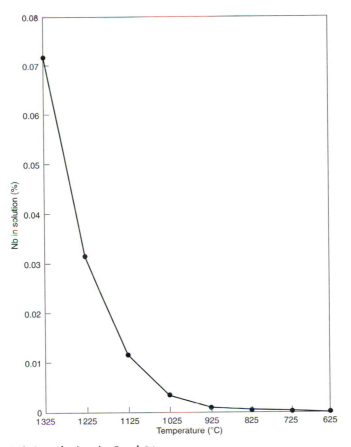

Fig. 13   % Nb in solution in Steel 91.

fore it is unlikely to be present after a normalise 1050°C + temper 750°C heat treatment. Furthermore, it is very highly alloyed with tungsten and molybdenum, Table 2(d), and thus is unlikely to form instantaneously at 600 and 650°C, but it does grow very quickly as indicated by its large size in Fig. 11. A composition reported in reference 10 for steel NF616, does not conform to that of Laves phase, Table 2(d).

The balance of the argument therefore suggests that the classical effect still holds for Steel 91 and its derivatives, as for 12% CrMoV steels,[11] that molybdenum and tungsten provide solid solution strength in addition to that given by the 9% Cr.

4.2 Precipitation Strengthening

The increased strength of Steel 91 over that of 9% Cr 1% Mo, which is of the order of 200% for 10 000–100 000 h at 600°C,[12] is due principally to the precipitation of VN and NbC during tempering in very small particle sizes, Figs

5–8. The compositions and sizes of these particles remain fairly constant over long periods at 600–700°C. The samples examined in this study were for Steel 91 material tested for up to 37 068 h at 650°C and 3000 h at 700°C, Table 2(c) and (d). Such durations, in parametric terms, are equivalent to 10^5–10^6 h at 600°C. This indicates that the particle compositions and sizes responsible for the precipitation strength will probably remain stable for very long periods at around 600°C which is the projected service temperature regime for Steel 91 and its derivative alloys.

The equivalent precipitates in the tungsten containing steels are similarly fine and have similar compositions.

There is no evidence to indicate, as some have suggested,[9] that the VN precipitates contain tungsten in the appropriate steels.

In both the tungsten free and tungsten added steels some precipitate particles have compositions approximately half way in composition between the VN Table 2(b) and NbC Table 2(c), as indicated in Table 2(e). VN and Nb(CN) are considered usually to have similar structures and lattice sizes and therefore might form contiguously as indicated in Figs 7 and 8. The evidence of Fig. 8 with the 'winged' particles is that VN has formed during testing in this case at 600°C on the NbC particles formed during tempering.

The role of tungsten in steels such as E911 and NF616 has been argued in this paper to be one of solid solution strengthening, which is in agreement with its recognised classic effect in steels such as 12% CrMoV.[11] However, others have argued for the existence of a strengthening effect from Laves phase[9,10] but this has not been proved with metallographic evidence. The Laves phase particles which have been observed in the studies reported here, and by others, are large and at or close to grain boundary positions, Fig. 11. The size and position of such particles are unlikely to contribute to the strength of the steel.

On the other hand they are probably detrimental to the properties of the steel, in terms of ductility, since Laves phase is recognised to be a brittle phase, behaving similarly to sigma phase in austenitic stainless steels, and also of strength, by removing tungsten and molybdenum in particular, from solid solution. Calculations and measurements have shown that a tungsten content of ~1.5% in solid solution, after normalising at ~1050°C, reduced to ~0.4–0.5% in solid solution after ~33 000 h at 600°C due to the formation of Laves phase and $M_{23}C_6$,[13] see Tables 2(a) and (d). The prognosis then is that the formation of Laves phase depletes the solid solution strength and therefore there will probably be a relatively faster decrease in strength in tungsten containing steels than for steels without tungsten over time at 600–700°C. Furthermore, since there will be an equilibrium amount of tungsten to remain in solution,[10,13] the greater the level of tungsten above this amount, the greater will be the loss in strength and ultimately the waste and cost of tungsten. It is for this reason amongst others that the Steel E911 is likely to be relatively more stable than other similar steels but which contain greater amounts of tungsten.

5. CONCLUSIONS

A study of the microstructure of Steel 91 and the equivalent Steel E911, containing 1% W, has shown that the long term high temperature strength of such steels derives from a basic strength largely determined by solid solution from the 9% Cr and 1% Mo base alloy with a small contribution from fine precipitates of NbC to which is added other components of strength. There is a significant component of strength from the precipitation of VN, occurring mainly during tempering, typically at 750°C, but also during subsequent testing at lower temperatures. This component can be maximised by having a composition close to stoichiometric for VN. The VN precipitates formed are smaller than 0.1 μm, and are stable in size over long periods at 600 and 650°C. The composition of these precipitates becomes closer to that of VN, from one containing initially significant amounts of chromium and some iron, with increasing time and temperature of testing. NbC particles are also formed which are much closer to the 'purer' form and therefore do not change as much in composition during testing.

The main precipitate type in terms of volume fraction is $M_{23}C_6$ which appears at prior austenite and martensite lath boundaries. At the latter position the $M_{23}C_6$ particles can be quite small, typically 0.2–0.5 μm and thus may contribute some dispersion strengthening.

In the tungsten containing steel, most is in solid solution after normalising and tempering with some in the $M_{23}C_6$ particles. During testing at 600–700°C, Laves phase, typically 50% W, 12% Mo, 30% Fe, 8% Cr, forms thus removing from solid solution tungsten and molybdenum in particular. The Laves phase forms as large particles, 0.5–1 μm, and thus offers no benefit to strength and may in fact cause deterioration in ductility parameters. It is considered that the Steel E911 will be more stable in strength than other grades containing higher amounts of tungsten.

It is anticipated that steels of the E911 type will provide a cost and technically effective solution to increasing demands from industry for power generation plant to operate at higher temperatures and pressures leading to greater thermal efficiency and more environmentally friendly operation.

ACKNOWLEDGEMENTS
The authors thank Mr. A.T. Sheridan (Acting Manager, Swinden Technology Centre) for permission to publish this paper.

REFERENCES
1. H.D. Newell: *Metals Progress*, Feb. 1936, 51.

2. M. Caubo and J. Mathonet: *Rév. Metall.*, 1969.

3. P. Patriarca, S.D. Harkness, J.M. Duke and L.R. Cooper: 'US Advanced Materials Development Program for Steam Generators', *Nuclear Technology*, March 1976, **28**, 516–536.

4. J. Orr and D. Burton: 'Development, Current and Future Use of Steel 91', *Ironmaking and Steelmaking*, 1993, **20**(5), 333.

5. R. Kishore, R.N. Singh, T.K. Sinha and Kashyap: 'The Morphology and Ageing Behaviour of δ-Ferrite in a Modified 9Cr-1Mo Steel', *J. Nuclear Materials*, 1992, **195**, 198–204.

6. J. Orr and D. Burton: 'Improving the Elevated Temperature Strength of Steel 91 (9% CrMoNbVN)', *Materials for Advanced Power Engineering 1994, Part 1*, Liege, 3–6 October 1994, published by Kluwer Academic Publishers.

7. K. Tokuno, K. Hamada and T. Takeda: 'Dispersion Strengthening High Cr Steels by forming V-Wings', *J. Metals,* 1992, April, 25–28.

8. J. Orr and D. Burton: 'A Study of the Basic Constitution of 9-11% Cr Steels for Elevated Temperature Service', *Draft ECSC Final Report*, British Steel Ref. FR S242–10 921 (ECSC Agreement No. 7210.KF/804).

9. Y. Tsuchida, K. Okamoto and Y. Tokunaga: 'Improvement of Creep Rupture Strength of High Cr Ferritic Steel by Additions of W', *ISIJ Int.*, 1995, **35**(3), 317–323.

10. J. Hald: 'New Steels for Advanced Plant up to 620°C', *Material Comparisons Between NF616, HCM12A and TB12M – III: Microstructural Stability and Ageing Conference*, E. Metcalfe ed., EPRI/National Power, London, 11 May 1995.

11. J.Z. Briggs and T.D. Parker: *The Super 12% Cr Steels*, Climax Molybdenum Company, 1965.

12. *PD 6526 Part 1: 1990 + AMD–7977: Elevated Temperature Properties for Steels for Pressure Purposes, Part 1, Stress Rupture Properties*, BSI Standards, London.

13. H. Cerjak, V. Foldyna, P. Hofer and B. Schaffernak: 'Microstructure of Advanced High Chromium Boiler Tube Steels', *this publication*.

Development of Advanced High Chromium Ferritic Steels

V. FOLDYNA, Z. KUBOŇ, A. JAKOBOVÁ AND V. VODÁREK
Vítkovice, a.s. Research and Development
Pohraniční 31, CZ-706 02 Ostrava 6, Czech Republic

ABSTRACT

Main strengthening and degradation mechanisms of Cr steels and the influence of chemical composition and microstructure are discussed. Solubility and precipitation of VN and Nb(C,N) in austenite and ferrite is analysed as well as the kinetics of Laves phase precipitation. Microstructure of P91 steel after short-term creep exposure at 600°C showed that inside subgrains small MX particles are precipitated on dislocations and at the same time Laves phase containing silicon is detected. A possible mechanism of strengthening in boron bearing chromium steels is discussed.

Larson-Miller parametric equation is used to elucidate the influence of C_{LM} and the data set on the reliability of creep strength estimation. Reliable extrapolation of creep properties is possible only in stress and temperature domain in which only one creep and creep rupture mechanisms operate.

1. INTRODUCTION

The creep properties of modified chromium steels are controlled by the chemical composition of these steels and their microstructure. Given a certain chemical composition, their microstructure depends on the heat treatment.[1]

The fundamental knowledge of strengthening mechanisms and microstructural evolutions during creep exposure is a prerequisite for successful development of chromium steels with higher creep resistance. Creep behaviour of these steels responds very sensitively to the structural changes during creep and makes it extremely difficult to predict creep properties using relatively short-term creep rupture tests over 10^4 to 2.10^4 hours. In the past four decades, much effort was devoted to unsuccessful attempts to develop more creep resistant materials, because short-time creep rupture tests showed ostensibly promising creep properties.[3–6] The really promising ways how to improve the creep resistance of the well-known steels X12CrMo 9.1 (9%Cr–1%Mo) and X20CrMOV 12.1 (12%Cr–1%Mo–0.3%V) were found about 20 years ago. These ways were based on increasing nitrogen content and modifying with low niobium content (~0.1 wt.%). This development led to the grade 91 steel. Further development was focused in the improvement

73

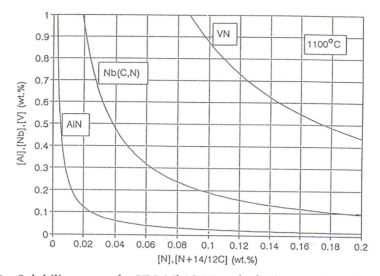

Fig. 1   Solubility curves for VN, Nb(C,N) and AlN in austenite with respect to 11 wt.% Cr in the steel.

of creep resistance of grade 19 by increasing Mo or Mo equivalent (%Mo + 0.5%W) in the steel, respectively by adding boron.[8–11] Similar steels with higher chromium content were developed in order to improve oxidation resistance.[2,12]

The aim of this paper is to analyse the principal strengthening mechanisms and degradation processes which influence the creep resistance. Special attention is also paid to creep rupture strength assessment with respect to structural changes occurring during creep exposure.

2. MICROSTRUCTURE, STRENGTHENING AND DEGRADATION MECHANISMS

2.1 Microstructure

Modified 9–12% chromium steels are commonly used in normalised or quenched and tempered condition. Austenising treatment dissolves the majority of nitrides, carbides or carbonitrides. The only secondary phase staying undissolved at austenising temperature is Nb(C,N). Solubility curves of relevant secondary phases can be used for assessment the conditions of dissolution of precipitates. The solubility curves for VN, Nb(C,N) and AlN at 1100°C are shown in Fig. 1 for a steel containing 11 wt.% chromium. As the solubility of VN is significantly higher than these for AlN and Nb(C,N), complete dissolution of VN can be expected at relatively low temperatures. In steels with reasonably low Al contents, AlN dissolves completely during normalising, too. On the other hand, Nb(C,N) particles are desirable to re-

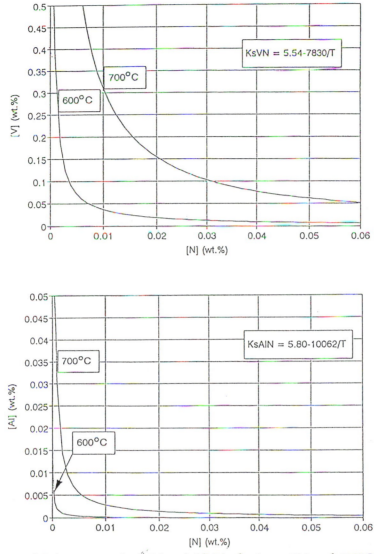

Fig. 2 Solubility curves for VN and AlN in ferrite at 700 and 600°C with respect to 11 wt.% Cr in the steel.

strict austenite grain growth. In the normalised or quenched condition chromium modified steels usually have martensitic structure without delta ferrite. After tempering the dominant secondary phase is $M_{23}C_6$ carbide, which particles are preferentially situated on the grain and subgrain boundaries or on the martensitic laths. The volume fraction of $M_{23}C_6$ is governed by carbon content of the steel. During tempering or creep exposure precipitation of VN and/or Nb(C,N) takes place due to lower solubility of these phases in ferrite. Moreover, precipitation of AlN takes place, too.

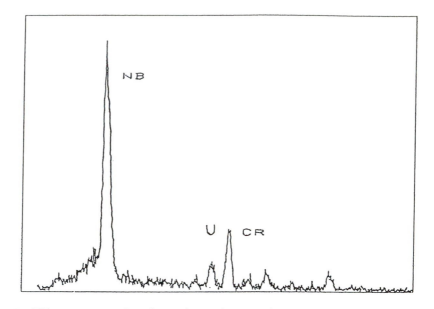

Fig. 3 ED spectrum taken from the niobium rich MX particle.

The Solubility curves of VN and AlN in ferrite at 700 and 600°C are shown in Fig. 2. Due to still high solubility product of VN at 700°C it can be assumed that vanadium nitride will precipitate mainly during creep at 600°C. The extent of precipitation of VN depends on nitrogen content in the solid solution after normalising or quenching (so called available nitrogen). Vanadium carbide was also sometimes expected to precipitate in 9 to 12% Cr steels. However, there is significantly greater solubility product for VC compared to VN. From the thermodynamical point of view, the precipitation of VC in mentioned chromium steels with low vanadium content is very improbable.

Typical ED spectra for two different MX phases are shown in Figs 3 and 4. In Fig. 3 there is Nb rich MX phase with small amounts of Cr and V, in Fig. 4 there is V rich MX phase with small amounts of Cr and Nb. Electron-energy-loss spectrometry (EELS) is more sensitive to light elements where conventional EDS is limited. A typical EELS spectrum of MX phase is shown in Fig. 5. The EELS analysis of vanadium rich particles of MX phase has confirmed that this phase is rich in nitrogen.[13] Niobium particles were not analyzed by EELS. Except of coarser particles of Nb(C,N) which most probably remained undissolved during normalising there were also observed small Nb rich precipitates situated on the subgrain boundaries and even in subgrain interior. They seem to precipitate during tempering or even during creep exposure and it can be expected that the amount of these fine particles will increase with increasing cooling rate from the normalising temperature. Fast

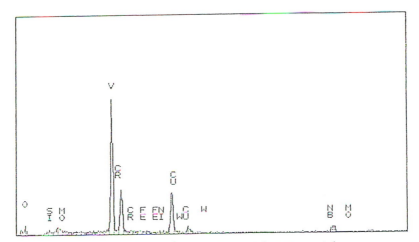

Fig. 4   ED spectrum taken from the vanadium rich MX particle.

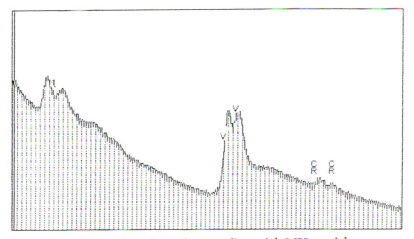

Fig. 5   EELS spectrum taken from the vanadium rich MX particle.

kinetics of Laves phase was confirmed by the presence of this phase in microstructure of steel P91 after creep exposure at 600°C for 957 hours (Fig. 6). This phase contained relatively high amount of silicon (the total Si content in the steel was 0.43 wt.%). At the same conditions, high density of MX particles was observed inside the subgrain interior (Fig. 7).

2.2 Strengthening and Degradation Mechanism

The principal strengthening mechanisms in chromium modified steels are solid solution and precipitation strengthening. The precipitation strengthening (PS) of low nitrogen containing steels (e.g. X12CrMo9.1,

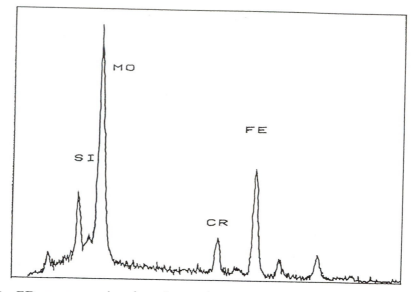

Fig. 6   ED spectrum taken from Laves phase.

X20CrMoV12.1) is effected predominantly by $M_{23}C_6$ carbides. With decreasing interparticle spacing (IPS) of $M_{23}C_6$ the proof stress at room temperature increases, while the creep rate decreases. In chromium steels modified with V and/or Nb the IPS further decreases due to precipitation of VN and/or Nb(C,N). Vanadium nitride precipitating especially on dislocations inside subgrains during creep exposure plays the decisive role in the excellent creep resistance of newly developed Cr-steels with higher nitrogen content.[10,14,15] Moreover, precipitation of Laves phase $Fe_2(W,Mo)$ can contribute to PS of these steels, at least in the first stage of creep exposure. When assessing PS in steel P91, Tokuno *et al.*[16] have considered only contribution of VN and Nb(C,N). Nevertheless, contribution of other phases is not negligible. It should be taken into account, that relatively large particles ($M_{23}C_6$, Laves phase) are located mainly on subgrain boundaries, but much smaller particles of VN and Nb(C,N) are situated especially inside subgrains. Therefore, it seems to be reasonable to estimate separately the IPS particles situated on subgrain boundaries (ℓ_{sgb}) and inside subgrains (ℓ_{sg}). The effective IPS (ℓ_{eff}) can be calculated as follows:

$$\frac{1}{\ell_{eff}} = \frac{1}{\ell_{sgb}} + \frac{1}{\ell_{sg}} \tag{1}$$

It was frequently supposed that significant solid solution strengthening

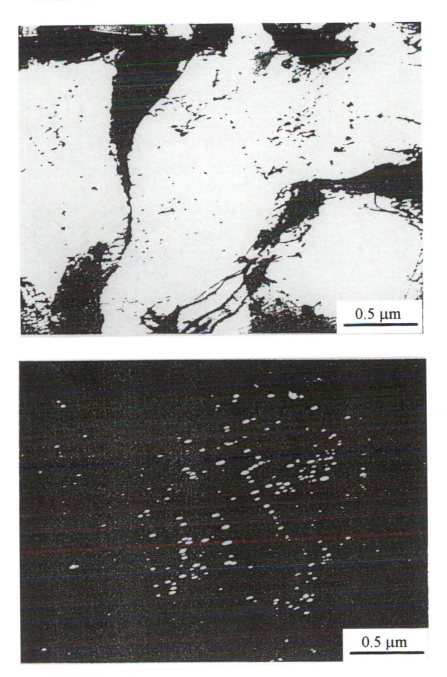

Fig. 7 Microstructure of P91 steel after creep exposure at 600°C for 957 hours:
a) bright field image
b) dark field image taken using $(200)_{MX}$ reflection

(SSS) of modified Cr steels can be attained by increasing Mo and/or W content in the steel.[1] Nevertheless, when assessing the SSS we must take into account the decreasing content of these elements in solid solution (SS) in the course of creep exposure. Hald[17,18] and Mimura et al.[19] described the changes in the amount of W content in secondary phases ($M_{23}C_6$ and Laves phase) during creep exposure of the steel Nf616 (9%Cr–0.5%Mo–1.8%W,V,Nb). They showed that the amount of precipitated W reached its quasiequilibrium state in roughly 20 000 hours at 600°C, in 6000 h at 650°C and in 2000 h at 700°C. At 600°C the remaining W content in SS is less than one third of the total W content in the steel. In this case, the strengthening effect of W is primarily caused by the precipitation of $Fe_2(W,Mo)$ during the creep process and not by SSS.[10,18] Afterwards, coarsening of $Fe_2(W,Mo)$ will act as one of the most effective degradation processes as the coarsening rate of Laves phase is much higher than that of $M_{23}C_6$.[10,15]

It is commonly accepted that the higher creep resistance of the new generation of chromium modified steel is a result of extensive precipitation strengthening by VN and Nb(C,N). The decisive role in PS by these particles is due to the nitrogen content in the steel. As a measure of the extent of PS, the nitrogen content in solid solution at working temperature -N_{SS}- can be considered in the simple form:

$$N_{ss/WT} = N_{ss/AT} - N_{AlN/WT} \qquad (2)$$

where WT and AT mean working and austenising temperatures.

The analysis made on 33 heats of chromium modified steels covering 9Cr–1Mo, 9Cr–1Mo(V), 12Cr–1Mo–0.3V, 12Cr–1Mo–1W–0.3V and 9Cr–1Mo–0.2V–0.05Nb–0.05N yielded in the equation correlating creep rupture strength at 600°C and 10^5 hours with proof stress at room temperature and nitrogen content in solid solution in the form:

$$R_{mT/100\,000\,h/600°C} = a + b \cdot R_{p0.2} + c \cdot N_{ss} \qquad (3)$$

Here a, b and c are temperature dependent regression quantities, $R_{p0.2}$ is the proof stress at room temperature and N_{ss} represents the amount of N in solid solution that is not bound as AlN and/or Nb(C,N) and/or TiN. The product $c \cdot N_{ss}$ determines the contribution of VN to the creep rupture strength.[10,15]

The analysis how aluminium and niobium contents can affect the amount of N_{ss} and therefore also creep rupture strength was made on the model steel with the following chemical composition (in wt.%): 0.15C; 10.5Cr; 1Mo; 0.2V; 0.04 to 0.50Nb; 0.005 to 0.045Al and 0.030 to 0.070N. These combinations of chemical composition cover most of typical chemical compositions for rotors, blades and bolts but they are quite applicable for tubes or plates too.

Addition of aluminium deteriorates creep resistance of chromium modified steels. AlN precipitating in steel lowers nitrogen available for VN precipi-

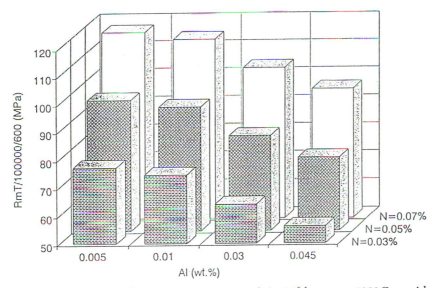

Fig. 8 Dependence of creep rupture strength in 10^5 hours at 600°C on Al and N contents of the model chromium steel.

tation, at the same time coarse AlN particles situated on grain boundaries lower creep ductility of the steel.[13,20]

Niobium addition up to 0.10 wt.% is sufficient to restrict the austenite grain growth. More Nb in the steel results in more Nb(C,N) particles staying undissolved during normalising. These particles have only limited influence on PS of the steel because they are relatively coarse.

Figures 8 and 9 indicate how the creep rupture strength in 100 000 h at 600°C depends on Al, Nb and N contents in the model steel with 0.07Nb, 0.05N and 0.01Al with the proof stress at room temperature 650 MPa. Given identical Nb content 0.07 wt.% (Fig. 8) or identical Al content 0.01 wt.% (Fig. 9) increasing N content from 0.03 to 0.07 wt.% leads to increasing creep rupture strength of about 60%. Lowering Al content from 0.045 to 0.005 wt.% produces higher N content in solid solution (N_{ss}) and the creep rupture strength increases of about 30% (Fig. 8). In the considered amounts Nb has only a slight effect on the creep rupture strength. Lowering niobium content from 0.1 to 0.04 wt.% leads to increasing creep rupture strength only of about 3.5% (Fig. 9).[15] In Fig. 10 the significant drop in creep rupture strength is observed with increasing Nb content in the steel.[10,13]

A good correspondence was observed between calculated and measured creep rupture strength of Cr-modified steel for bolts with high Nb content confirming the detrimental effect of high niobium content.[20,21]

So far, a not fully understood strengthening mechanism operates in boron bearing steels. Good creep rupture properties were reported for the

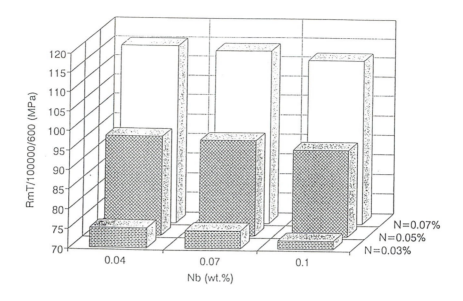

Fig. 9   Dependence of creep rupture strength in 10^5 hours at 600°C on Nb and N contents of the model chromium steel.

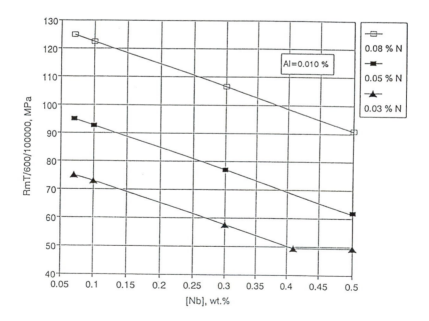

Fig. 10   How the creep rupture strength of the model chromium steel depends on Nb and N contents in the steel.

0.17C–9.36Cr–1.55Mo–0.27V–0.06Nb–0.015N steel with 0.010 wt.%B. As nitrogen content in this steel is relatively small, the observed creep resistance cannot be explained by the strengthening contribution of VN. It is commonly accepted that the improved creep rupture behaviour of boron bearing steels is associated with the replacement of carbon atoms in $M_{23}C_6$ carbides by boron to form $M_{23}(C,B)_6$ with high resistance against coarsening. Formation of $M_{23}(C,B)_6$ was experimentally confirmed by using atom-probe-field-ion microscopy (APFIM),[22] but the effect of boron on coarsening rate of $M_{23}(C,B)_6$ is questionable. Lundin et al. proposed another explanation for good creep properties of this type of steels, so called latent creep resistance.[22] According to this model precipitation process occurs successively during creep in a dynamic process. In this case, a delicate balance between nucleation and dissolution of small precipitates is suggested. The precipitates nucleate on dislocations taking advantage of the strain field around the dislocation core, thus effectively pinning the dislocations. When the dislocation manages to break away, the precipitate becomes unstable and dissolves. This process can be repeated and should lead to a decrease in creep rate during steady state creep. In steels with sufficiently high boron content boron can probably segregate to dislocations and enable nucleation of small particles of VN on dislocations. High dislocation density is of course desirable. It is reasonable to expect that this mechanism will be more efficient in steels with lower nitrogen content not allowing boron nitride to precipitate but sufficient for some precipitation of VN. Very fine VN particles, repeatedly nucleating and dissolving can act as obstacles for dislocation movement and may reduce the rate of recovery as well. This seems to be also acceptable explanation for high dislocation density after long-term creep exposure in low nitrogen boron-bearing steel X17CrMoVNbB9.1.[21]

2.3 Kinetics of Laves Phase Precipitation in P91 Steel

It was proved experimentally that even in chromium modified steels with Mo or Mo equivalent contents about or less than 1 wt.% Laves phase precipitates during creep exposure.[7,23,24] For assessing the contribution of Mo and/or W to the solid solution strengthening it is necessary to determine the part of these elements staying in solid solution at equilibrium condition when all Laves phase is precipitated. As this precipitation is diffusion controlled, the kinetics of this process can be described by Johnson-Mehl-Avrami equation.[25] Construction of the kinetic curves for Laves phase precipitation was made for Nf616 steel by Hald.[17,18] It was showed that kinetics of $Fe_2(Mo,W)$ in Nf616 was very fast and after approximately 20 000 hours at 600°C the amount of W in secondary phases reached its equilibrium amount.

A similar approach for determining Laves phase kinetics was utilised in course of P91 steel. Kinetic curves were constructed for two heats with following chemical composition (in wt.%).

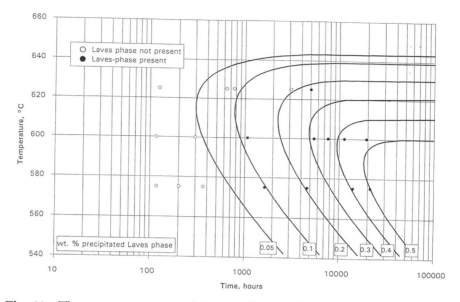

Fig. 11 Time-temperature precipitation diagram for Laves phase in the steel P91, heat B.

Heat A:
0.09C–0.54Mn–0.23Si–8.32Mo–0.46Ni–0.20V–0.06Nb–0.063N–0.014Al

Heat B:
0.10C–0.40Mn–0.41Si–8.48Mo–0.10Ni–0.22V–0.09Nb–0.043N–0.005Al

For calculating the equilibrium amount of Mo in Laves phase the software for thermodynamic calculations 'Thermocalc' developed at the Swedish Royal Institute of Technology was used similarly as in Nf616 steel.[17,18] Calculated solution temperatures of Laves phase in steel A and B were below 500°C, which is considerably lower than that observed in P91 steel.[7] The probable explanation of this discrepancy are incorrect thermodynamic data for Mo-rich Laves phase incorporated into Thermocalc. Therefore, providing that $M_{23}C_6$ is stable and its composition changes very little with regard to Mo content, the calculated solution curve of Laves phase was shifted towards higher temperature to fit the experimental results. This procedure was described elsewhere.[26] Kinetic curves were then calculated as well as time-temperature precipitation diagrams – see Figs 11 and 12. It was observed that the steel with higher silicon content had faster kinetics of Laves phase precipitation and this phase dissolved at a slightly higher temperature, too. Silicon is known to promote Laves phase precipitation,[7] although the exact mechanism of its influence is not clear. When comparing Nf616 and P91 the kinetics of Laves phase precipitation seems to be similar in both steels. But the extent of

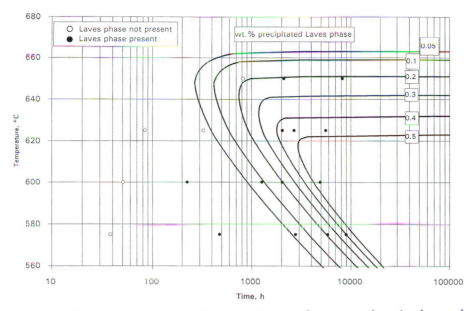

Fig. 12 Time-temperature precipitation diagram for Laves phase in the steel P91, heat B.

Laves phase precipitation in Nf 616 steel is much higher and the volume fraction of secondary phases was higher too.

3. CREEP DATA PREDICTION

According to the concept of the deformation and fracture mechanisms map we should accept the assumption that creep rupture data attained within the domain where creep or creep fracture is governed by one dominant mechanism cannot be used for prediction to another domain, where creep and fracture are controlled by different dominant mechanism. As soon as the condition is changed and the boundary is crossed into another domain,[27–29] the prediction becomes unreliable.

Two distinct domains of the stress dependence of creep rate and time to rupture have been frequently observed in the case of precipitation strengthened low alloy and modified Cr steels.[27,30] The values of stress exponent and apparent activation energy are shown in Table 1.

The creep rupture data have been analyzed by means of the equation as follows:

$$t_r = A \cdot \sigma^{-n_r} \cdot \exp\left(\frac{Q_r}{RT}\right) \qquad (4)$$

where t means time to rupture [h]; σ is applied stress [MPa]; n_r is stress expo-

Table 1 Mean values of stress exponent and apparent activation energy

Steel	Low stress domain		High stress domain	
	Q_r[kJ.mol^{-1}]	n_r	Q_r[kJ.mol^{-1}]	n_r
Low alloy CrMoV	315	−3.7	375	−7.6
9Cr−1Mo, 9Cr−1Mo−V, 12Cr−1Mo−V	500	−4.7	600	−10

Table 2 Activation parameters (Q_r, n_r), C_{LM} and creep rupture strength of the COST steel D3

External conditions	Q_r [kJ.mol^{-1}]	n_r	C_{LM}	Creep rupture strength in 10^5 h at 600°C [MPa]
Low stress domain $\sigma \geqslant \sigma_{z(T)}$	317	−3.5	14.4	74
High stress domain $\sigma \geqslant \sigma_{z(T)}$	605	−9.3	30.5	120

nent; Q_r represents apparent activation energy [kJ.mol^{-1}]; A is a constant and R, T have their usual meaning.

This equation was used for estimation of Q_r and n_r separately for low stress and high stress domains. The stress characterising the transition between these domains (break point $\sigma_{z(T)}$) was found to be closely related to the critical stress σ_{crit}, which can be expressed as

$$\sigma_{crit} = A^* . \frac{G.b}{l} \tag{5}$$

where constant $A^* \approx 1$; G is the shear modulus; b is the length of Burgers vector and l is the interparticle spacing of secondary phases.

The temperature dependence of $\sigma_{z(T)}$ is mainly effected by Ostwald ripening respectively additional precipitation of new particles during the creep process. In low alloy CrMoV steel it was observed that $\sigma_{z(T)} \leqslant \sigma_{crit}$, while in 9CrMoV steel was observed that $\sigma_{z(T)} \geqslant \sigma_{crit}$.[27,31] It is assumed that in low-alloy CrMoV steel the coarsening of V_4C_3 particles prevails in the course of creep process. On the other hand, in 9CrMoV steel the interparticle spacing decreases during creep due to precipitation of VN fine particles.[1,14,15] It is assumed that the break point corresponds to Orowan stress which is necessary for bowing dislocations between particles.

It is supposed that in the high stress region dislocations can overcome particles of secondary phases by Orowan mechanism, while in the low stress domain dislocation should climb over particles.[32] Dislocations can climb over obstacles in different ways.[33,34] Arzt and Ashby[35] have performed analysis how to estimate relevant threshold stress σ_T which is necessary for dislocation to be able to climb over particles. It was found that σ_T can be estimated by the relationship:

$$\sigma_T \approx k \cdot \frac{G \cdot b}{l} \tag{6}$$

where k varies from 0.3 to 0.04.

In dependence on local differences in microstructure, different deformation mechanisms were observed, even in one specimen of precipitation strengthened materials. Moreover, after structural changes occurring during creep exposure, changes in deformation mechanism should be expected. Creep behaviour of these steels responds very sensitively to the structural changes in the course of creep and makes it extremely difficult to predict their creep properties using relatively short term creep or creep rupture tests, over 10^4 to 2.10^4 hours.[2,27–29] Greenfield has quoted examples of creep rupture strength prediction for 12CrMoVNb steel based on data up to 10^4 hours. Larson-Miller parametric equation (constant C = 25) led to over-optimistic prediction. Later long-term tests revealed that the true creep rupture strength was only half the predicted figures.[2]

3.1 Application of Larson-Miller Equation

Very different procedures are proved using the well-known Larson-Miller parametric equation:

$$P_{LM} = f(\sigma) = T(C_{LM} + \log t_r) \tag{7}$$

where C_{LM} is a constant which original formulation was considered to be between 20 and 25; t_r is time to rupture. Nowadays, P_{LM} is used either with elected (and often sophisticated) C_{LM} constant ranging from 25 to 40 or with calculated C_{LM} constant.

Clearly, the greater magnitude of constant C_{LM} the more optimistic estimation of the creep rupture strength. We have tried to estimate the most convenient C_{LM} constant for assessment of creep rupture strength of modified Cr steels. We have used two procedures based on the following assumptions:

a) in the low stress domain (below the break point) as well as in the high stress domain (above this point), the stress dependence of the parameter may be assumed to be linear and can be expressed by:

$$P_{LM} = a_0 + a_1 \cdot \log \sigma \tag{8}$$

b) throughout the investigated temperature and stress interval, the stress dependence of

$$P_{LM} = a_0 = a_1 \cdot \log \sigma + a_2 \cdot (\log \sigma)^2 + \ldots + a_k \cdot (\log \sigma)^k \tag{9}$$

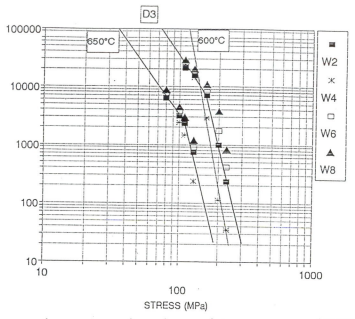

Fig. 13 Stress and temperature dependences of time to rupture of COST steel D3.

where $a_0, a_1, a_2 \ldots a_k$ are regression constants and k should be at least 3.

Using the first procedure we have found that in the high stress domain much higher C_{LM} (over 31) can be expected than that in the low stress domain (below 22). Using the second procedure the result depends on the relation between creep tests performed in the high stress domain and low stress domain. When most of tests is performed in the high (low) stress domain the calculated C_{LM} constant should be closer to the C_{LM} constant expected in the high (low) stress domain. Evidently, $C_{LM} \geq 31$ indicates that most of creep tests were performed in the high stress domain and more or less significant overestimation of assessed creep rupture strength values in 100 000 h can be expected. In this case, overestimation should be expected even though the attained time to rupture was relatively long, about 40 000 h.

In another paper it was shown[36] how increasing C_{LM} constant leads to an extraordinary overestimation of predicted time to rupture of the COST steel D3. In the mentioned still (0.16C–11.3Cr–1.8W–0.3Mo) it was found that the break point $\sigma_{z(T)}$ is about 160 MPa at 600°C and about 110 MPa at 650°C. Figure 13 shows the stress and temperature dependence of time to rupture calculated with aid of relevant L–M parametric equations valid for high and low stress domain. The ascertained activation parameters (Q_r, n_r), C_{LM} constants and creep rupture strengths are shown in Table 2. Steel D3 was tested after 4 different kinds of heat treatment but no significant differences in creep prop-

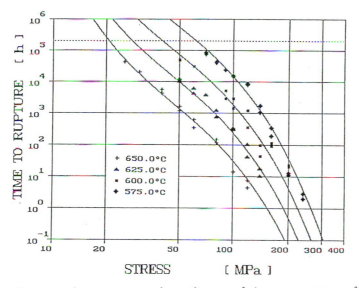

Fig. 14 Stress and temperature dependences of time to rupture of a 9%Cr–1%Mo–0.1%V steel calculated by means of the general parametric equation.

erties were observed. Only after heat treatment marked as W4 lower creep resistance was observed but only in the high stress region.[8] Our estimation based on the thermodynamical calculations and eqn (3) confirmed that there was no reason to expect significant differences in creep resistance of examined steel.

Nevertheless, creep resistance of steel D3 close to 100 MPa was expected. To explain the reason why the observed values were lower may be the unfavourable influence of nickel (0.79 wt.%) on creep resistance of steel D3.

In steels with molybdenum equivalent (Mo + 0.5 W) $\geqslant 1$ precipitation strengthening in high stress domain is further increased by Laves phase precipitation and the break point $\sigma_{z(T)}$ increases. Therefore, the high stress–low stress domain boundary is moved, too. It is evident, that sufficiently long-term creep rupture tests should be carried out for reasonable prediction of creep rupture strength.

It is believed that at low stresses the creep rupture strength can decrease to an 'inherent creep strength'[37] or 'matrix strength'[38] through a loss of precipitation and solid solution strengthening effect. So called sigmoidal behaviour characterised by abrupt decreasing of creep rupture strength due to loss of strengthening effects was observed more than 30 years ago.[39] As a general rule, the point of inflection shifts with increasing temperature to shorter times to rupture and more or less lower stress values. Sigmoidal behaviour seems to be common for all types of ferritic creep resistant steels. Nevertheless, number of creep rupture curves does not show any point of inflection up to 100 000 hours. It is assumed that for longer times or higher temperatures this behaviour can be expected.[39] Strang and co-workers have shown that in 12CrMoV and 12CrMoVNb steels the point of inflection shifted at given temperature to shorter times with increasing nickel content in these steels.[40,41] Figure 14 shows the stress and temperature dependencies of time to rupture of the steel 9%Cr–1%Mo–0.1%V. The points of inflection are at all testing temperatures at about 49 MPa. Sigmoidal behaviour is evident only at higher temperatures (625 to 650°C). Concerning the above mentioned steel D3, the possibility of sigmoidal behaviour at 600°c up to 100 000 hours is not eliminated. So we should admit that our estimation of creep rupture strength can be somewhat underestimated. There is evidence that sigmoidal behaviour can lead either to overestimation or to underestimation of predicted creep rupture strength. We should emphasize the importance of long-term creep tests performed at working as well as at higher temperatures.

4. CONCLUSION

Nitrogen content in solid solution N_{ss} seems to have the most significant effect on high creep resistance of chromium modified steels because it controls the extent of precipitation strengthening by fine VN particles.

With increasing niobium and aluminium contents creep rupture strength decreases due to reducing the available nitrogen content for VN precipitation.

So called 'latent creep resistance' seems to be acceptable explanation for observed unexpected high dislocation density after long-term creep exposure, as well as high creep resistance in the low nitrogen boron-bearing steel X17CrMoVNbB 9.1.

When assessing the creep rupture strength of advanced 9–12% Cr steels, structural changes occurring during creep should be taken into account. Precipitation of Laves phase increases the precipitation strengthening only in the first period of creep exposure. Afterwards, lowered contents of Mo and/or W in solid solution have a much lower solid solution strengthening effect than expected.

In the high stress domain we can expect the C_{LM} constant over 31, in the low stress domain below 22. The higher used constant C_{LM} the more optimistic estimation of the creep rupture strength. Sigmoidal behaviour makes the creep rupture strength more complicated. The importance of long-term creep rupture tests was emphasised.

ACKNOWLEDGEMENT

The authors would like to acknowledge the financial support from the Grant Agency of Czech Republic – Grant No. 106/94/0853 and the Ministry of Education of Czech Republic – Grant OC 501.30. At the same time COST 501 Management Committee is acknowledged too for financing one month's scientific mission of Mr. Kuboň at TU Lyngby.

REFERENCES

1. V. Foldyna, A. Jakobová, V. Vodárek and Z. Kuboň: *Proc. Conf. Materials for Advanced Power Engineering 1994*, C.R.M., Liége, Belgium, October 1994, 453.

2. P. Greenfield: *Proc. Conf. High Temperature Materials for Power Engineering 1990*, Kluwer Academic Publishers, Liége, Belgium, October 1990, 423.

3. G. Guntz: 'Ferritic tubes, pipes for high temperature use in boilers', *The T91 book*, Vallourec Industry, France, 1990.

4. M. Caubo: *Proc. Conf. Ferritic Steels for Fast Reactor Steam Generators*, British Nuclear Energy Society, London, 1978, 218.

5. G. Gunz, F. Pellicani, J. Hollis and B. Duquise: *ibid*, 164.

6. T. Yukitoshi , K. Yoshikawa, K. Tokimasa, T. Kudo, Y. Shida and Y. Jnaba: *ibid*, 87.

7. C.R. Brinkman, B.J. Gieseke, D.J. Alexander and P.J. Maziasz: *Proc. First International Symposium on Microstructure and Mechanical Properties of Aging Materials*, Chicago, USA, November 1992.

8. C. Berger, R.B. Scarlin, K.H. Mayer, D.V. Thornton and S.V. Beech: *Proc. Conf. Materials for Advanced Power Engineering 1994*, C.R.M., Liége, Belgium, October 1994, 453.

9. K. Hidaka, M. Shiga, S. Nakamura, Y. Fakui, N. Shimiau, R. Kaneko, Y. Walanabe and T. Fujifa.: *ibid*, 281.

10. V. Foldyna and Z. Kubon: *Proc. Conf. Materials Engineering in Turbines and Components*, Newcastle upon Tyne, UK, April 1995, 373.

11. K. Spiradek, R. Bauer and G. Zeiler: *Proc. Conf. Materials for Advanced Power Engineering 1994*, C.R.M., Liége, Belgium, October 1994, 251.

12. Y. Sawaragi, A. Jseda, K. Ogawa and F. Masuyama; in Proc. see [1], 309.

13. V. Voldárek: "Report of the short-term scientific mission at TU Tampere', COST 501/III, WP 11, November 1994.

14. V. Foldyna, A. Jakobová, R. Říman and A. Gemperle: *Steel Research*, **62**, 1990, 453.

15. Z. Kubon and V. Foldyna: *Steel Research,* in press.

16. K. Tokuno, K. Hamada and T. Takeda: *JOM*, 1992, **44**, **25**.

17. H. Hald: *Proc. 6th EPRI-RPI 403–50 Meeting*, London, UK, June 1993.

18. H. Hald: *Proc. the EPRI National Conference New Steels for Advanced Plant up to 620 C*, London, UK, May 1995, 152.

19. H. Mimura, M. Ohgami, N. Naoi and T. Fujita: *Proc. of Materials for Advanced Power Engineering Conference 1994*, C.R.M. Liége, Belgium, October 1994, 361.

20. K. König: *Proc. 16 Vortragsveranstaltung Langzeitverhalten warmfester Stähle und Hochtemperaturwerkstoffe*, VDEh Düsseldorf, November 1993, 45.

21. Z. Kuboň and V. Foldyna: 'Consideration of the Role of Nb, Al and Trace Elements in Creep Resistance and Embrittlement Susceptibility of 9–12 % Cr Steel', *Performance of Bolting Materials in High Temperature Plant Applications*, York, UK, 1994, Institute of Materials, 1995, 175.

22. L. Lundin, S. Fällman and H.-O. Andrén: *Microstructure and Mechanical Properties of a 10% Chromium Steel with Improved Creep Resistance at 600 C*, to be published.

23. A. Iseha, H. Ternaishi and K. Yoshikava: *Tetsu-to-Hagané*, 1985, **71**, S1341.

24. F. Brühl, H. Cerjak, P. Schwaab and H. Weber: *Steel research* 1991, **62**, 75.

25. 'Steel – A Handbook for Materials Research and Engineering', *Vol. 1 – Fundamentals*, VdEh Düsseldorf, 1992, 94–100.

26. Z. Kuboň: 'Report of the short-term scientific mission at TU Lyngby', *COST 501.III*, WP 11, April 1994.

27. V. Foldyna: 'Creep of low alloy and modified chromium steels' (In Czech), *Technické aktuality VÍTKOVIC*, 1988, No. 1.

28. V. Foldyna, A. Jakobová and V. Kupka: *Proc. VIIIth International Symposium on Creep Resistant Metallic Materials*, September 1991, Zlín, Czechoslovakia, 186.

29. V. Foldyna, A. Jakobová and V. Kupka: *Proc. Conf. Creep and Fracture of Engineering Materials and Structures*, B. Wilshire and R.W. Ewans eds, The Institute of Metals, London, UK, April 1993, 573.

30. V. Foldyna and J. Purmenský: *Czechoslovak Journal of Physics*, 1989, **B39**, 1133.

31. V. Foldyna, A. Jakobová and Z. Kuboň: 'Assessment of Creep Resistance of 9–12 % Cr Steels with Respect to Strengthening and Degradation Processes', *Materials Aging and Component Life Extension*, Milano, Italy, October 1995.

32. L.M. Brown and R.K. Ham: *Strengthening Methods in Crystals*, A. Kelly and R.B. Nicholson eds, Applied Sci. Publ., London, 1971, 1.

33. R.S.W. Shewfelt and L.M. Brown: *Phil. Mag.*, **30**, 1974, 1135.

34. R.S.W. Shewfelt and L.M. Brown: *Phil. Mag.*, **35**, 1977, 945.

35. E. Arzt and M.F. Ashby: *Scripta Met.*, **16**, 1982, 1285.

36. H. Cerjak, V. Foldyna, P. Hofer and B. Schaffernak: This proceedings volume.

37. K. Kimura, H. Kushima, K. Yaqi and C. Tanaka: *Proc. Conf. Creep and Fracture of Engineering Materials and Structures*, B. Wilshire and R.W. Evans eds, The Institute of Materials, U.K., April 1993, 555.

38. H. Chikwanda, M. McLean and A. Strang: see [2], 291.

39. J.H. Bennewitz: *Proc. Joint International Conference on Creep*, New York-London, 1963, 69.

40. A. Wickens, A. Strang and G. Oakes: *Journal of Institute of Mechanical Engineering*, 1980, 11.

41. A Strang: private communication.

Microstructural Development in Advanced 9–12%Cr Creep Resisting Steels – a Collaborative Investigation in Cost 501/3 WP11

R.W. VANSTONE,[*] H. CERJAK,[**] V. FOLDYNA,[***] J. HALD,[#]
AND K. SPIRADEK[##]

[*] GEC ALSTHOM Turbine Generators, Rugby, UK
[**] Technical University of Graz, Institute of Materials, Graz, Austria
[***] Vitkovice j.s.c. Research Institute, Ostrava, Czech Republic
[#] ELSAM/ELKRAFT, Lyngby, Denmark
[##] Austrian Research Centre, Seibersdorf, Austria

ABSTRACT

The progress of collaborative activity in COST 501/3 WP11 to determine and to model the microstructural evolution and mechanisms of creep resistance of advanced 9–12%Cr steels is described. The first phase of the investigation yielded a qualitative description of the way in which these microstructures evolve during creep. The second phase is yielding quantitative data on the dispersion and chemical analysis of the phases present and on the evolution of the dislocation structure. Analysis of these quantitative data has begun and has already revealed those microstructural parameters which correlate with creep rupture strength. Different parameters are significant in determining rupture strength at different temperatures and after different lifetimes.

1. BACKGROUND

Since the early 80s there has been extensive activity within Europe with the objective of the development of a new generation of high creep strength 9–12%Cr steels for application in power generation plant. In particular much effort has been concentrated on the development of rotor forgings and castings for steam turbines. Most of this activity has taken place under the auspices of COST 501 and has involved collaboration between major European turbine-makers, foundries, forgemasters, utilities and research institutes.

In 1981 work began under Round 1 of COST 501 on the development of boron containing steels.[1] The success of this work encouraged its continuation in Round 2, beginning in 1987, when the scope of the work was also widened to investigate steels containing increased additions of nitrogen, molybdenum and tungsten (Table 1).[2] These steels were produced as small

Table 1 COST 501/2 programme, chemical composition (%) of test melts

Steels		C	Si	Mn	Cr	Mo	Ni	V	W	Nb	B	N
Rotor forging steels												
A	+Nitrogen	0.05-0.08	0.1-0.3	0.2-0.7	9-12.5	1-2	0.5-1.0	~0.2	–	~0.06	–	0.1-0.3
B	+Boron	0.15-0.18	~0.1	~0.1	9-10.5	~1.5	~0.1	~0.3	–	~0.06	0.005-0.01	–
D	+Tungsten	0.10-0.16	~0.1	~0.5	10-12	<0.5	0.5-1.0	~0.2	~2.0	~0.06	–	<0.07
E	+Tungsten/ Molybdenum	0.10-0.18	~0.1	~0.5	10-12	~1	0.5-1.0	~0.2	~1.0	~0.06	–	<0.07
F	+Molybdenum	0.10-0.18	~0.1	~0.45	9.5-12	1-2	0.5-1.0	~0.2	–	~0.06	–	<0.07
Casting steels												
C	+Chromium	~0.13	~0.3	~0.6	~10.5	~1	~0.8	~0.2	–	~0.07	–	~0.05
CT	+Tungsten	~0.13	~0.3	~0.6	~10.0	~1	~0.8	~0.2	~1	~0.07	–	~0.05

scale melts and investigated in a number of heat treatment conditions in order to assess the effect of different solution treatment temperatures and different levels of proof strength produced by different tempering temperatures. All the steels developed for application to high temperature rotor forgings were cooled at a controlled rate of $120°C\,hour^{-1}$ after the solution treatment in order to simulate the cooling rate at the centre of a large rotor forging. The steels were investigated by tensile, impact, and creep tests and their tensile and impact properties after long term ageing were also determined. Creep data of over 30 000 hours are now available on these steels. The results of these investigations have been very successful. Steels have been identified with sufficient creep strength for application to steam turbine rotors and castings operating at temperatures of up to 600°C. The toughness and long term stability of these steels has also been demonstrated to be more than adequate.

The success of these investigations was such as to justify the manufacture of full scale components, rotor forgings and valve chests, in the most promising steels. Samples taken from these components have been investigated to show that short term properties fulfil the expectations derived from the earlier work[3, 4] and longer term tests are currently in progress as part of the activity of Round 3 of COST 501.

Although the investigation outlined above has clearly been very successful it has provided only an empirical understanding of the properties of these steels. In order to provide a more fundamental understanding of the microstructures and mechanisms responsible for the enhanced properties, and in particular the enhanced creep strength, of these steels a major metallographic study of the steels investigated is necessary. The acquisition of such understanding is expected to facilitate the development of these steels still further thus enabling even greater advances in the operating conditions of steam tur-

Table 2 Participants in microstructural characterisation and modelling activities

Participant	Location
Austrian Research Centre	Seibersdorf, Austria
Technical University of Graz, Institute of Materials	Graz, Austria
Institute of Physics of Materials (Czech Academy of Sciences)	Brno, Czech Republic
Skoda Research	Plzen, Czech Republic
Vitkovice j.s.c. Research Institute	Ostrava, Czech Republic
ELSAM/ELKRAFT	Lyngby, Denmark
Tampere University of Technology, Institute of Materials Science	Tampere, Finland
VTT Metals Laboratory	Espoo, Finland
ENEL Thermal and Nuclear Research Centre	Milano, Italy
University of Ancona Dept. of Mechanics	Ancona, Italy
Swedish Institute for Metals Research	Stockholm, Sweden
Cambridge University, Dept. of Materials Science and Metallurgy	Cambridge, United Kingdom
GEC ALSTHOM Turbine Generators	Rugby, United Kingdom

bines. A great strength of the collaborative nature of work carried out under COST is that the resources necessary for a study of this nature can be made available. Under Round 3 of COST 501, which began in January 1993, thirteen organisations from seven different countries are involved in detailed microstructural investigations (Table 2) and this activity is being co-ordinated by GEC ALSTHOM.

In parallel with these metallographic activities, mathematical models are being developed to describe the evolution of the microstructure and properties of these steels. This includes the application of thermodynamic models to predict equilibrium phases.

These activities represent an attempt to apply metallographic characterisation to the explanation of mechanical test results in order to provide a powerful tool for further alloy development. The work will maximise the value gained from the extensive mechanical testing carried out in previous rounds of COST 501. The ultimate goal is the prediction of alloys and microstructures with optimum properties, especially high creep strength, leading to the further development of 9–12%Cr steels for high temperature applications.

2. MATERIALS INVESTIGATED

The study is currently focusing on microstructural characterisation of the small scale melts tested in Round 2. Material from these investigations is available in the as heat treated condition and after long term exposure to stress and elevated temperature. Materials are available from the most successful melts and also from the less successful steels. Examination of these materials enables the description of microstructural evolution under stress and elevated temperature and will allow the identification of those characteristics respon-

Table 3 Steels tested in Round 2 selected for metallographic investigation.

Austenitising temp (°C)	1020		1070		1100–1120	
0.2% Proof Strength (MPa)	≥600	≥700	≥600	≥700	≥600	≥700
Rotor forging steels						
Steel A					√	√
Steel B					√	
Steel D	√	√	√	√	√	√
Steel E	√		√	√	√	
Steel F			√			
Casting steels						
Steel C					√	
Steel CT					√	

√ Investigated by creep tests in Round 2
√ Investigated by metallography in Round 3

sible for the improved performance of one steel with respect to another. Despite the large metallographic resource available for this investigation it is still not possible to study all the materials available from earlier work. Table 3 illustrates all the different steels and different heat treatment conditions tested in Round 2. It also indicates those steels which have been selected for microstructural characterisation. The selected conditions allow comparison to be made between different steels given similar heat treatments and between the same steel in different conditions of heat treatment. Steel A was not selected for investigation because it showed insufficient hardenability for application to large rotor forgings.

The tests in Round 2 involved creep tests at 600 and 650°C. The microstructural investigations currently in progress are characterising ruptured creep specimens tested at these temperatures and comparing them with the as heat treated condition. Examination of both the head and gauge length of the creep specimens is being carried out to enable the effect of creep strain in the gauge length to be established in comparison with the unstrained material in the head of the specimen.

3. MICROSTRUCTURAL CHARACTERISATION
Microstructural characterisation of each of these conditions and locations involves determination of the following parameters:

– hardness

– grain size

– primary phase description:

   – phase (martensite, delta ferrite etc.)

   – chemical composition

– secondary phase description:

   – species ($M_{23}C_6$, MX, Laves phases etc.)

   – size distribution

   – location (intra-granular, cell boundaries etc.)

   – chemical composition

   – inter-particle spacing

– dislocation structure:

   – distribution (random, sub-cells etc.)

   – cell size and shape

   – dislocation density

Determination of these parameters is being carried out according to agreed common procedures involving the use of optical microscopy, scanning and transmission electron microscopy, electron diffraction, energy dispersive spectroscopy, residual phase analysis and automated image analysis. The first phase of the investigation, now largely complete, has involved establishment of a qualitative description of these parameters. The second phase of the investigation, which is currently in progress, involves their quantitative description.

4. TRENDS IN MICROSTRUCTURAL EVOLUTION

The qualitative phase of the investigation has led to the observation of several general trends.

The hardness of all steels was lower in the creep specimens than in the as heat treated condition. Softening appears to be accelerated by creep strain so that the gauge length of creep specimens was softer than the head. Softening occurred more rapidly at 650°C than at 600°C.

The grain size of the wrought steels examined varied from 50 to 400 microns. As would be expected, higher solution treatment temperatures gave larger grain sizes. Grain size in the cast steels was much larger, being in the order of millimetres.

With the exception of the casting steel, CT, all steels consisted of 100% tempered martensite. In steel CT about 3% delta ferrite was present.

$M_{23}C_6$ carbides were present in all steels in all conditions. In the as heat treated condition typical particle size was in the range of 50 to 150 nm. In the

Fig. 1a   As heat treated condition × 22 500

Fig. 1b   Gauge length of creep specimen after 13 000 hours at 600°C × 15 000

Fig. 1   Evolution of microstructure in steel E.

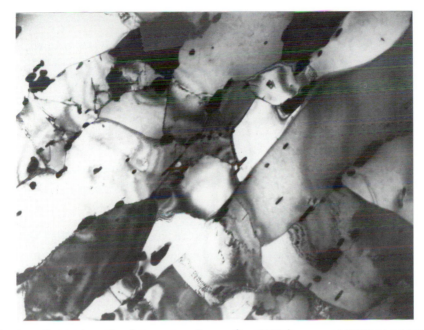

Fig. 1c   Gauge length of creep specimen after 9000 hours at 650°C × 11 000

Fig. 1   Evolution of microstructure in Steel E

creep specimens they were significantly coarser, particularly in the gauge length. These particles were rich in Fe, Cr, Mo and W and in several cases became further enriched in these elements after creep. Most of these particles were associated with grain or sub-grain boundaries.

Although sometimes difficult to identify, MX particles were observed in all of the steels with the exception of steel D. They were generally much finer than the $M_{23}C_6$ particles, their size typically ranging from 20 to 50 nm, and they were rich in V and Nb. They were relatively unaffected by the creep process, either coarsening very slightly or in one case appearing to go in to solution. Unlike $M_{23}C_6$ these particles were found mainly inside sub-grains rather than at the boundaries.

M_2X particles were also sometimes difficult to identify but they were observed in a few cases. They appeared to show a tendency to dissolve after exposure to creep conditions.

Laves phases were observed in the delta ferrite of the cast steel CT in the as heat treated condition but were initially absent in all the wrought steels. After creep, Laves phases appeared in nearly all the steels. However in the tungsten-free steel, steel F, their appearance was delayed for several thousand hours at 600°C and they were not observed at all in the creep specimen tested at 650°C. These phases generally appeared as fairly large particles, typically about 500 nm in size, on grain and sub-grain boundaries. Particles were enriched in Fe, W, Cr, and Mo.

In all the steels dislocation re-ordering and annihilation occured during creep. This led to an initial reduction in dislocation cell-size, as dislocations re-ordered to form new sub-grains, and then to growth of these sub-grains. These changes were accompanied by a reduction in dislocation density. These processes occured more rapidly in the gauge length of creep specimens than in the head and led to the formation of larger sub-grains. Steel B has been observed to be noticeably more resistant to this dislocation re-ordering than many of the other steels.

Micrographs derived from steel E are presented in Fig. 1 to illustrate the way in which microstructure is typically seen to evolve.

5. ANALYSIS OF QUANTITATIVE MICROSTRUCTURAL DATA

In order to ensure that results from different laboratories are comparable, each laboratory has characterised a common reference material. Results derived from this reference material have revealed significant differences in the values of microstructural parameters reported by different laboratories. The origin of these discrepancies has been investigated by additional characterisation of a set of reference micrographs and a set of micrograph tracings. The results indicate that significant differences arise from the application of different image analysis techniques. This work has enabled normalisation of results from different laboratories to take account of differing metallographic and image analysis techniques and thus to allow the comparison of results from different sources.

Analysis of the quantitative data is now in progress. Early work has concentrated on establishing correlations between microstructural parameters in the as-received condition and creep strength. As Table 1 indicates, data are available from steel E in several heat treatment conditions. For this steel the microstructural parameters which correlate with rupture strength have been identified. In the short term (<50 000 hours at 600°C and <4000 hours at 650°C) rupture strength correlates directly with proof strength and with the following microstructural parameters:

$$\sigma_r = f(L_{min}^{-1}, \rho_{dn}, d_e, V_{dens,} IPS^{-1}, Cr^{-1}, Nb^{-1}, V^{-1}) \quad \text{where:}$$

$$\sigma_r = \text{rupture strength}$$
$$L_{min} = \text{dislocation cell width}$$
$$\rho_{dn} = \text{dislocation density}$$
$$d_e = \text{mean particle size}$$

$$V_{dens} = \text{particle volume density}$$

$$IPS = \text{inter-particle spacing}$$

$$Cr, Nb, V = \text{solid solution content of Cr, Nb, V}$$

Although it must be recognised that many of the parameters in this relationship are inter-dependent, some physical interpretation of the relationship is possible. The relationship is consistent with creep resistance being associated with all those features of the as-received microstructure which inhibit dislocation movement. A dense dislocation network and particle dispersion lead to high creep resistance. The inverse relationship between creep strength and concentration of elements in solid solution suggests that these elements give much more effective resistance to creep strain by forming particles than by remaining in solid solution.

At longer times (>50 000 hours at 600°C and >4000 hours at 650°C) the correlations between rupture strength and proof strength and many of these microstructural parameters break down:

$$\sigma_r = f(P_{wt\%}, Cr^{-1}, W^{-1}, Mo^{-1}, V^{-1}) \quad \text{where: } P_{wt\%} = \text{total weight \% of isolated particles}$$

$$W, Mo = \text{solid solution content of W, Mo}$$

There is now reduced dependence of rupture strength on as-received microstructure. However it still appears that creep resistance is favoured by formation of particles rather than by solid solution strengthening. The inverse influence of the solid solution content of W and Mo, not noted as an influence on shorter term rupture strength, is particularly significant. It is likely that this reflects the role played by Laves phase precipitation in this steel. Further investigations to explain these observations are planned.

6. MICROSTRUCTURAL MODELLING

Several participants in this COST activity are actively pursuing the development of mathematical models to describe the behaviour of these steels. The quantitative data currently being generated will provide a valuable reference against which these models may be tested and the results generated to date already indicate the most important parameters to be modelled.

Thermodynamic computer packages already exist for the prediction of equilibrium microstructures and these are being applied to explain, for example, the influence of tungsten level on the evolution of Laves phases. Kinetic models are also essential to a complete understanding and these are also being developed, first of all to predict microstructure in the as heat treated condition and then to predict how this microstructure develops during creep.

The interaction between the parallel activities of microstructural modelling and microstructural characterisation enables both activities to be focused on the parameters of most significance.

7. CONCLUSION

The metallographic investigations and data analyses completed to date have already made a valuable contribution to our understanding of the alloys investigated and their microstructural evolution. Although the materials investigated were developed with steam turbine applications in mind, the fundamental understanding of the behaviour of these steels can be applied to materials for other equally important components for power generation, such as those in the boiler, and also to materials for high temperature components in other industries. The continued generation of quantitative data and its analysis and assimilation through the development of microstructural models will enable refinement of that understanding. This in turn will facilitate the further development and optimisation of 9–12%Cr steels, allowing their application at even higher temperatures and thus the realisation of more efficient power generation.

ACKNOWLEDGEMENT

Acknowledgement is made to National Power for financial support of the work by Cambridge University and to the financial support of the Commission of the European Union and of those national governments who have provided support.

REFERENCES

1. B. Walser and G.H. Gessinger: 'Material developments for advanced coal-fired power plants in the COST project', *Proc. 1st EPRI Int. Conf. 'Improved Coal-Fired Power Plants'*, Palo Alto, California. November 1986.
2. B. Scarlin and P. Schepp: 'State of the European COST activities on improved coal-fired power plant', *Proc. 2nd EPRI Int. Conf. 'Improved Coal-Fired Power Plants'*, Palo Alto, California. November 1988.
3. C. Berger, R.B. Scarlin, K.H. Mayer, D.V. Thornton and S.M. Beech: 'Steam Turbine Materials: High Temperature Forgings', *Proc. COST 501 Conf. 'Materials for Advanced Power Engineering'*, Liege, October 1994.

4. R.B. Scarlin, C. Berger, K.H. Mayer, D.V. Thornton and S.M. Beech: 'Steam Turbine Materials: High Temperature Castings', *ibid*.

Evolution of Microstructure and Properties of 10% Cr Steel Castings

K.H. MAYER,* H. CERJAK,** P. HOFER,** E. LETOFSKY,**
F. SCHUSTER***

* MAN Energy GmbH., Nürnberg, Germany
** Technical University of Graz, Dept. of Materials Science and Welding Technology,
Graz, Austria
*** VOEST Alpine Stahl Linz – Foundry, Linz, Austria

ABSTRACT

There are strong environmental and economic pressures to increase the thermal
efficiency of fossil fuel fired power stations, and this has led to a steady increase
in steam temperatures and pressures resulting in world wide plans for ultra-
supercritical power plants. Based on present developments, it can be assumed
that, with the improved ferritic steels on the basis of 9–10%CrMoVNbN, the
present limit for the steam admission temperature of 565°C can be raised to approximately 600°C. Under the European joint programme COST 501–2, a tung-
sten-alloyed 10% CrMoVNbN version, subjected to modified heat treatment,
has been identified as a very promising cast steel. The castability and weldability of which is comparable to that of traditional cast steels 1%CrMoV (GS–17
CrMoV 5 11). After optimisation of chemical composition and foundry technique the commercial production was started successfully.

INTRODUCTION

World-wide environmental concerns such as saving our resources and the
need to reduce CO_2 emission have become a challenge. An improvement in
power plant efficiency can be achieved by increasing the operating temperature and pressure of steam turbines. This improvement is dependent on the
development of improved higher temperature materials of construction.

Steel castings play a key role in highly-loaded turbine components, such as
outer and inner casings, valve bodies, steam inlet pipes, elbows, etc.[1] For use
up to 600°C improved 9–12% Cr-steel castings, are required. Cast materials
with ferritic/martensitic microstructures can be used because of their
favourable physical properties such as high thermal conductivity and low coefficient of thermal expansion, coupled with higher resistance to thermal
shock. These are some of the advantages relative to austenitic stainless steels.

Some general aspects of castability and weldability of the W-modified cast
steel G-X 12 CrMoWVNbN 10 1 1 are given in this paper.

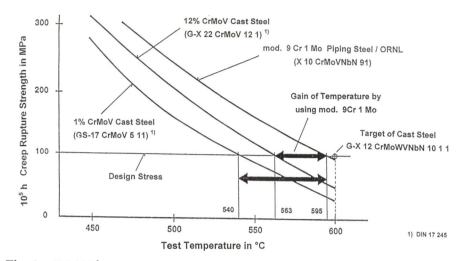

Fig. 1   100 000 h creep rupture strength of ferritic cast and wrought steels.

MATERIAL DEVELOPMENT

Property Profile

The desired properties of the cast steel were defined as follows:

- A 100 000 h creep strength of 100 MPa at 600°C.
- Good castability and weldability.
- Through-hardening capability up to about 500 mm wall thickness.
- Properties such as fracture toughness, low-cycle fatigue strength and long-term toughness corresponding at least to those of the ferritic cast steels currently used up to 565°C.

Figure 1 shows the 100 000 h creep strength versus the test temperature of the modified 9% Cr 1% Mo steel, developed in US and standardised as i.e. as P91 in ASTM A335 in comparison with the traditional cast steels 1% CrMoV, German designations GS-17 CrMoV 5 11 and 12% CrMoV (G-X 22 CrMoV 12) used in Europe for temperatures up to 565°C.[2,3]

Within the framework of the European COST 501 programme (Development of materials for advanced steam cycles[4]) a tungsten alloyed 10% Cr cast steel, German designation G-X 12 CrMoWVNbN 10 1 1, was designed to fulfil the increased demands on the creep strength for advanced design fossil fired power plants and components.

DEVELOPMENT OF CAST STEELS IN COST 501 ROUND 2

Pre-Evaluation Programme

Based on the results of earlier tests and the evaluation of the literature test

Table 1 Chemical composition of the test plates, the valve body and an IP-inner casing

Chemistry of Induction furnace heat no. 60 504–1 (in weight-%):
Test Plate – G-X 12 CrMoVNbN 10 1

C	Si	Mn	P	S	Al	Cr	Ni	Mo	W	V	Nb	N
0.13	0.42	0.58	0.009	0.004	0.007	10.60	0.87	1.02	–	0.21	0.07	0.05

Chemistry of Induction furnace heat no. 60 504–2 (in weight-%):
Test Plate – G-X 12 CrMoWVNbN 10 1 1

C	Si	Mn	P	S	Al	Cr	Ni	Mo	W	V	Nb	N
0.13	0.33	0.52	0.012	0.004	0.006	10.50	0.86	1.03	1.01	0.23	0.066	0.049

Chemistry of AOD heat no. 81374 (in weight): Valve Body – G-X 12 CrMoWVNbN 10 1 1

C	Si	Mn	P	S	Al	Cr	Ni	Mo	W	V	Nb	N
0.12	0.29	0.62	0.027	0.003	0.013	10.51	0.93	0.99	0.99	0.22	0.08	0.048

Chemistry of LD-process heat no. 200930 (in weight %):
IP-Inner Casing – G-X 12 CrMoWVNbN 10 1 1

C	Si	Mn	P	S	Al	Cr	Ni	Mo	W	V	Nb	N
0.12	0.23	0.97	0.012	0.0023	0.010	9.45	0.86	0.98	1.03	0.21	0.06	0.049

melts (heats no. 60 504–1 and 60 504–2 of Table 1) were chosen for testing. Compared with the target values of mod. 9Cr1Mo according to ASTM 387, Grade 91, the Cr content was increased from 8.0–9.5% to 10.0–10.5% to improve the solubility of nitrogen and to obtain a ferrite-free martensitic microstructure. For the addition of tungsten a value of roughly 1% was chosen (heat no. 60 504–2).

Plates with the dimensions 800 × 400 × 100 mm were casted to check the test parameters. After preliminary tests, the heat treatment version B, BO, C and CO, according to Fig. 2, were chosen from the pre-evaluation tests in order to perform the long-term creep rupture tests. The test programme included the investigation of the strength and toughness properties as a function of the macro- and microstructure.

Results of Pre-Evaluation Programme
Strength and toughness properties. The 0.2% proof strength of the tungsten-based version is in the same region than the 0.2% proof strength for the 1% CrMoV and 12% CrMoV cast steels specified by DIN 17 245. The toughness properties for the tungsten-free version is characterised by a higher notch impact energy than the version alloyed with 1% tungsten. However, the toughness of the tungsten based version is also distinctly higher than the minimum impact energy specified by DIN 17 245 for the 1% CrMoV and 12% CrMoV

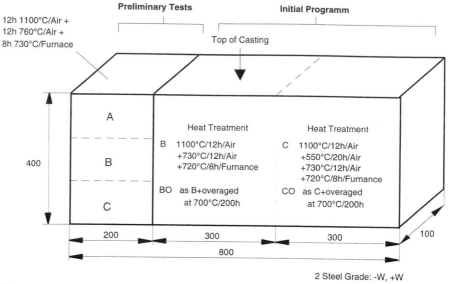

Heat No.: 60 504-1 (G-X 12 CrMoVNbN 10 1)   Heat No.: 60 504-2 (G-X 12 CrMoWVNbN 10 1 1)

Fig. 2 Cast steel material of evaluation test (test plates).

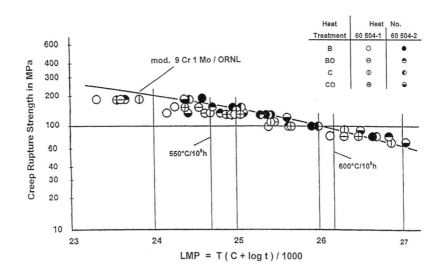

Fig. 3 Creep rupture strength of 10% CrMoVNbN (heat no: 60 504–1) and 10% CrMoWVNbN (heat no: 60 504–2) – cast steel, [heat treatment codes B, BO, C and CO are of Fig. 2].

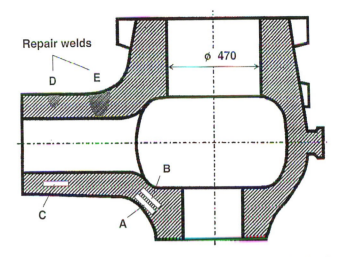

Fig. 4   Cast steel material of componenten programme (valve body)

cast steels, traditionally used in the manufacture of steam turbines. The results of the exposure tests carried out on the Charpy V-notched specimens at 480, 600 and 650°C up to 10 000 h shown a maximum reduction is at 600°C of the notch impact energy determined at room temperature.[2]

Creep tests. The creep tests of the pre-evaluation tests at 600 and 600°C had reached about 33 000 h. Figure 3 provides general details on the creep rupture strength (Larson-Miller diagram). The mean-value curve of the mod. 9Cr1Mo pipe steel provides a basis for comparison. The tungsten-containing version (heat no. 60 504–2) shows a higher creep strength over the full test period compared with the tungsten-free version and in the long term shows a slight superiority over the mod. 9Cr1Mo pipe steel for all four heat treatment conditions.

For the tungsten-free version (heat no. 60 504–1) the heat treatment condition C with the 550°C pretempering treatment features the highest creep strength whereas heat treatment condition B features the lowest creep strength. For all heat treatment conditions the tungsten-free version generally reflects a flatter pattern compared with the pipe steel which has a similar chemical composition. This is probably attributable to the longer heat treatment period required for castings which general results in increased and coarser carbide precipitation.

Prototype Valve Body
Based on the trial melts and the screening programmes for selecting the best chemical composition and heat treatment[5] a pilot valve body was casted to verify castability, non destructive inspectability and weldability.

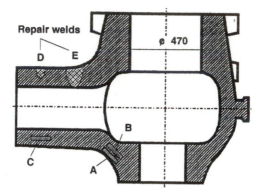

Location	0,2 – Limit MPa	Tensile Strength MPa	Elongation A 5 %	Reduction of Area %	Impact Energy Joule	FATT (50) °C
A	571	734	18.7	47	31	+ 63
B	567	719	20.2	43	30	+ 60
C	586	742	18.2	45	30	+ 45
D　transvers	–	758	–	51 [1]	–	–
weldmetal	–	–	–	–	27	–
E　transvers	–	616 [2]	–	8 [1][2]	–	–
weldmetal	744	854	15.4	51	31	+ 63

1) fracture base metal
2) micro shrinkage

Fig. 5 Mechanical properties of pilot valve body, G–X 12 CrMoWVNbN 10 1 1 (heat no: 81 374).

The composition being chosen for the pilot valve body was given in Table 1. For the investigation a total of 5 different specimen positions (see Fig. 4) have been chosen – one rim and two core zones as well as two manufacturing weld zones. The number and size of the defected flaws were similar to those found in 1% CrMoV cast steel (GS–17 CrMoV 5 11) traditionally used in the manufacture of steam turbines.

Results of the Valve Body
The mechanical properties determined at positions A to E of the valve body are shown in Fig. 5. The specimen positions A and B represent the thickest wall sections, whereas the specimen position C shows the profile of properties in the thin-walled support. The properties of manufacturing welds are determined at the positions D and E.

The results of exposure tests, carried out at 480, 600 and 650°C up to

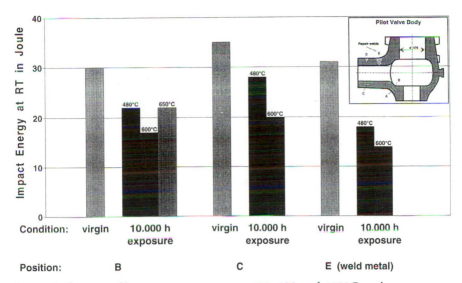

Fig. 6 Influence of long term exposure at 480, 600 and 650°C on impact energy at RT of cast steel G-X 12 CrMoWVNbN 10 1 1 (valve body).

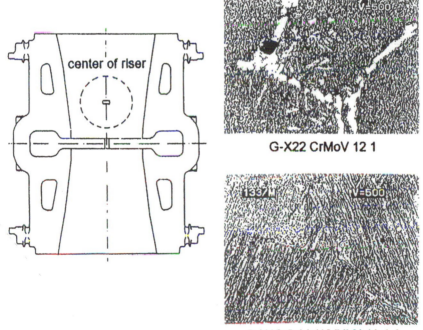

Fig. 7 Micro structure in heavy section of inner casing.

Fig. 8. IP-inner casing with welded on cast-pipes (weight: 44 t).

10 000 h on the Charpy V toughness at room temperature are shown for position B and C of the valve body and for the weld metal, position E, in Fig. 6.

FOUNDRY PRODUCTION IN COMMERCIAL SCALE[6]

Metallurgical Aspects
The physical metallurgy of simply alloyed 10% Cr steels are well known and documented.[7] The system becomes more complicated by adding elements for example Mo and/or W which are beneficial for solid solution strengthening but may produce delta ferrite, unless balanced by addition of an austenite former that has the effect of lowering the required tempering temperature.

Delta ferrite is an undesirable phase in 10% Cr steels and has a detrimental effect on creep resistance, ductility and a strong influence in impact energy. The cooling rate during solidification has also an influence on quantity and shape of delta ferrite phase. In heavy sections of large steel castings more delta ferrite was observed than in smaller castings with thinner wall thicknesses. Unfortunately it is very difficult to get no delta ferrite formation when the Cr level is on the upper side of the specification range. Compared with the pilot valve casting (heat no. 81374, Table 1) the Cr content was reduced from 10.5

to 9.5%. Typical chemical analysis of the normal commercial production are listed in Table 1. Heat no. 200 930 presents the optimised composition for the IP-inner casing (see Fig. 8).

The complex alloy content of modified 10% Cr steels makes them inherently susceptible to segregations. For this reason also the other creep strength improving elements such as V, Nb, H have to be adjusted very precisely to their optimum quantity.

Heavy primary grain boundary carbides have been observed in large steel castings made of G-X 22 CrMoV 12 1 in the past which made them unacceptable. A typical microstructure of primary precipitated carbides in G-X 22 CrMoV 12 1 is shown in Fig. 7. Due to the lower cooling rates in sand molds and the high heat capacity of big risers which are necessary for feeding purposes, the carbon segregation could not be avoided if the carbon content of the liquid steel is in the range from 0.20 to 0.25% as specified for G-X 22 CrMoV 12 1.

The reduction of carbon from 0.22 to 0.12% C in the new designed high creep strength 9–12% Cr steels was also required to maintain good weldability.

Foundry Experience

In Fig. 8 one of the largest steel castings made using the new tungsten alloyed 10% Cr steel can be seen. This inner casting has a weight of 40 t and the two inlet pipes were separately casted and welded to the casting.

The solidification characteristics of the complex alloyed 10% Cr steel are different relative to the low alloyed 1% CrMoV steels. Risers and mold taper have to be increased for better feeding. Nevertheless the feeding technology was adjusted in heavy sections of large steel castings, after cutting off the risers, small interdendritic shrinkage and gas holes were observed. These local casting defects were removed and repaired by welding. Besides these observations, no other typical casting defects were found.

MICROSTRUCTURAL DEVELOPMENT DURING AGEING

A microstructural characterisation of the as received condition and of broken long-term creep rupture specimens, tested at 600°C and 650°C, of the martensitic cast steel G-X 12 CrMoWVNbN 10 1 1 (heat no.: 60 504–2; Code: CT) was performed. The microstructure was qualitatively and quantitatively investigated by hardness test, light and electron microscopic examination, including EDX investigation. The investigations were performed on three broken creep rupture specimens and a material in the virgin condition. The creep rupture testing was carried out at 600°C and 650°C under constant initial stresses of 80, 110, 130 MPa. The times to rupture were determined to be 7 959 h, 12 119 h and 33 410 h respectively.

The microstructure in light microscopy appears as a highly tempered martensitic structure with fine precipitates and vestiges of δ-ferrite decorated

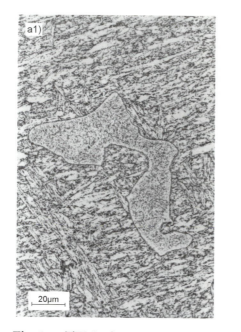

Fig. 9a   TEM microscopy:
a1) as received δ-ferrite, a2) as received;
G-X12 CrMoWVNbN 10 1 1 (heat no: 60 504–2, code: CT)

with fine precipitates. Figure 9 shows the TEM microstructure after ageing and creep. The martensitic phase of the as received material consists of lath with first signs of subgrains between the laths containing a high density of dislocations. It should be noted that this density of dislocations produced by martensitic transformation during quenching still remains after tempering. Precipitates with an average size of about 0.1 μm are distributed preferentially along lath boundaries, and appear with middling density. The microstructure of the creep rupture samples appears distinctly different to that of as received condition and is in its turn different in the head and shank region. The head portions show a decrease in dislocation density and the beginning of regularly grouped dislocations to first dislocation-free subgrains.

The shank portions represent a distinctly more recovered microstructure compared to that of the head portions. The dislocations are grouped to well defined oriented subgrains with quite low dislocation density inside them. Both the head and the shank show a heavy precipitation of coarsend particles located preferentially on subgrain boundaries.

The recovery behaviour of the head portions of the investigated samples shows an unexpected tendency. The head portion of specimen crept at 600°C, 33 410 h showing a subgrain structure rather than a lath structure is

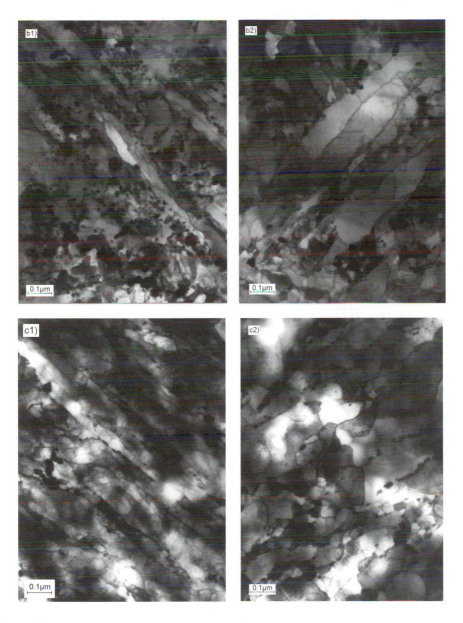

Fig. 9b   TEM microscopy:
b1) aged at 600°C, 33 410 h, b2) crept at 600°C, 33 410 h, 110 MPa;
c1) aged at 650°C, 7959 h c2) crept at 650°C, 7959 h, 80 MPa
G-X 12 CrMoWVNbN 10 1 1 (heat no: 60 504–2, code: CT)

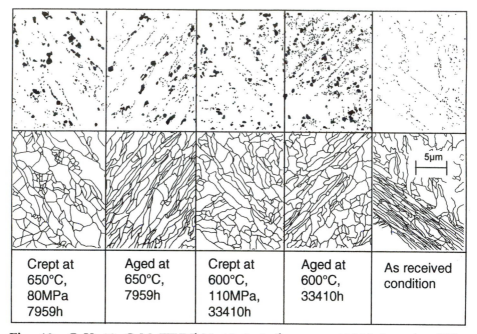

| Crept at 650°C, 80MPa 7959h | Aged at 650°C, 7959h | Crept at 600°C, 110MPa, 33410h | Aged at 600°C, 33410h | As received condition |

Fig. 10 G-X 12 CrMoWVNbN 10 1 1 (heat no: 60504–2, code: CT). Evaluation of particles and subgrains during creep and ageing.

distinctly more recovered than the head portion of sample exposed at 650°C, 7959 h which would have been expected to show greater recovery due to the much higher ageing temperature.

In order to extract the information of the distribution and size of particles and the general tendency of martensite lath width and its development during creep and ageing from the TEM microscopy a manual evaluation method was undertaken.[8] The evolution of both the particles and the martensite lath subgrains in the microstructure of the seven investigated conditions are separately shown in Fig. 10. The coarseness of particles and the growth of martensite lath subgrains can be observed to occur with the aid of temperature (600°C and 650°C) and beyond that with the aid of stress (130 MPa and 80 MPa).

Considering Fig. 10 it is quite obvious that there is a significant increase in the subgrain size with time and temperature in the head and the shank region in relation to the virgin condition. Owing to the high stress level in the shank the microstructure of this region is more recovered that that of the head area indicated by a significant difference in the martensite lath or subgrain width and dislocation density between head and shank region. Whereas the average distribution of the particle size is comparable at low magnifications (Fig. 10).

From this occurrence we can deduce the assumption that temperature (important for shank and head area) may mainly influence coarseness of particles.

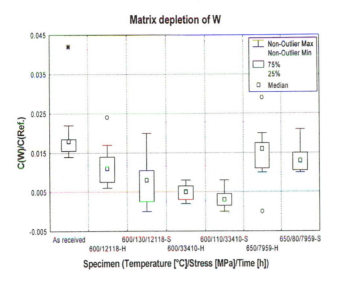

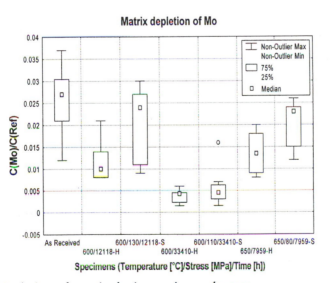

Fig. 11   Depletion of matrix during ageing and creep.

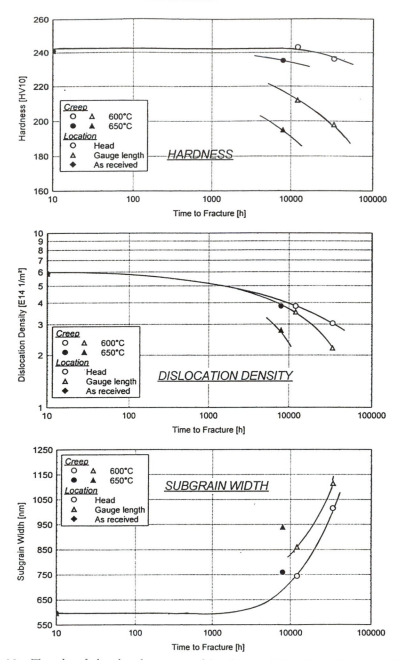

Fig. 12 Trends of the development of hardness, dislocation density and sub-grain size of G-X 12 CrMoWVNbN 10 1 1 as a function of temperature, time and stress.

Whereas stress (only important in the shank area) mainly contributes to the growth of martensite lath and the formation and growth of subgrains and the decease of dislocation density.

The microstructural changes of the creep rupture samples during the exposure at different times and temperatures were determined. Whereas the heads of the samples (H) were exposed to the temperature only, the shanks (S) were additionally crept by the load applied. The results of the microstructural changes, determined, i.e. by the amount of the W and Mo content, solved in the matrix, are given in Fig. 11. From these, it can be seen, that a depletion of W and Mo content in the matrix as a function of the exposure time and -temperature, as well as of the head or shank position were observed. In accordance to extended electron microscopical investigations and mathematical modelling of the status of equilibrium using Thermocalc-Programme[9] this observed depletion is caused by precipitation and strong coarsening of W and Mo rich Laves Phases and coarsening of $M_{23}C_6$ carbides, combined by a recovery process which is accelerated in the crept shank portion of the specimens (Fig. 10).[10]

Figure 12 shows the trends of hardness, dislocation density and sub-grain size of the martensitic cast steel G-X 12 CrMoWVNbN 10 1 1 as a function of temperature, time and stress using overage values, experimentally observed.[11] It is obvious that there is a significant decrease of hardness in the shank region of the creep samples relatively to the virgin condition as well as to the head region. The dislocation density decrease at longer times to fracture. The shank region shows a greater reduction than the head region. At longer times to fracture the sub-grain size increase characteristically relatively to the virgin condition. This result is in comparison with the results of Fig. 10. At higher testing temperature the diagrams show the same trends but at shorter times.

BASIC INVESTIGATIONS OF WELDABILITY

For the quantification of the influence of welding on the microstructure of G-X 12 CrMoWVNbN 10 1 1 Gleeble simulation, representing the manual metal arc welding process, were applied to produce HAZ-simulated microstructures. They were exposed to different PWHT-treatments and tested using hardness tests, metallographic investigations, constant strain rate tests, creep tests and toughness tests. Primary attention was given to the softening effect in the HAZ and its influence on the creep resistance of the welded materials. The decrease shown by the W-modified version seems to be less pronounced than that observed in the P91 material.

In Fig. 13, the results of constant strain rate tests applied on weld thermal cycle simulated material G-X 12 CrMoWVNbN 10 1 1 and P91 (seamless pipe, 149 mm outside diameter and 20 mm wall thickness, heat no. 856240) are shown. A minimum of creep resistance in the region exposed to the peak tem-

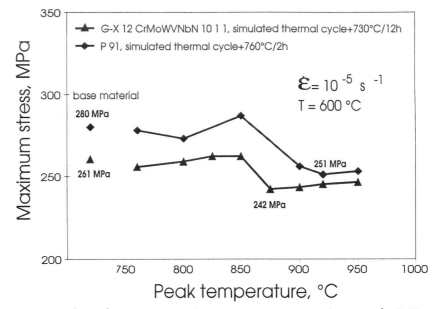

Fig. 13 Results of constant strain rate tests on specimens of G-X 12 CrMoWVNbN 10 1 1 and P91 subjected to weld thermal cycle simulation treatment followed by tempering (HAZ-softening).

perature about 920°C, that represents microstructure shortly above AC_1 can be observed. This is an explanation for the Type-IV cracking behaviour of welded compositions of this type of steel during long term creep exposure.[12,13]

Creep rupture tests on samples which were designed to represent material containing microstructures caused by a peak temperature of 920°C were performed. The results are shown in Fig. 14 and are compared with results obtained on creep samples taken from uninfluenced base material. As can be seen from Fig. 14 the creep resistance of HAZ simulated material falls remarkably below the creep resistance of the uninfluenced base material for the material P91. Comparing the behaviour of the material P91 to the material G-X 12 CrMoWVNbN 10 1 1, it can be revealed that the softening effect in the casting W-modified material on the creep behaviour is lower than that of material P91. These results still need to be confirmed by future investigations, especially by the testing of original welded and stress relieved samples and by intensive metallographic investigations. These tests are currently in progress and the results will be presented at a later date.

CONCLUSIONS

The cast steel G-X 12 CrMoWVNbN 10 1 1, developed and tested with the COST 501 Round 2 Programme, appears to have a creep strength which is at least as high as that of the modified 9% CrMo steel P91. Due to the solidifi-

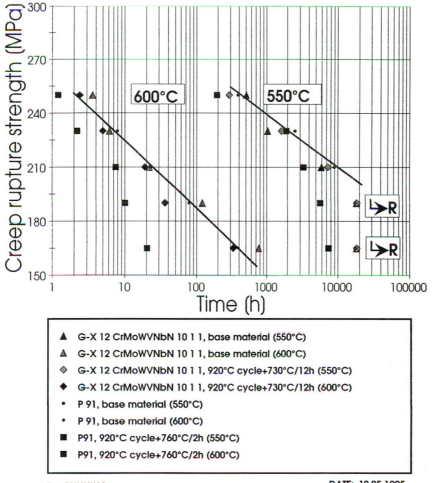

Fig. 14   Results of creep rupture tests of P91 and G-X 12 CrMoWVNbN 10 1 1 (heat no: 81374) in the as received and HAZ simulated condition.

cation characteristics of the 9–10% Cr cast steel, the castibility is different to the 1% CrMoV castings. Higher expenditures for feedings and padding additions are required for a good quality performance. The weldability and suitability for non-destructive testing are similar to that of the 1% CrMoV cast steel GS-17 CrMoV 5 11, conventionally used for turbine manufacturing. The toughness properties are also similar to those of this cast steel. The long-term tests will be continued within the framework of the third round of COST 501, WP11, in order to determine the service-relevant properties for plants operating at 600°C.

REFERENCES

1. F. Schuster and G. Köfler: *Berg- und Hüttenmännische Monatshefte*, 1982, Jahrgang 127, Heft 1, 1–6.

2. R.B. Scarlin, C. Berger, K.H. Mayer, D.V. Thornton and S.M. Beech: *Materials for advanced power engineering 1994, Part 1*, D. Coutsouradis *et al.* eds, Netherlands, Kluwer Academic Publishers, 73–88.

3. K.H. Mayer, W. Gysel and A. Trautwein: 'Modified 9% CrMo cast steel for castings in improved coal-fired power plants', *Proc. Conf. Improved coal-fired power plants*, San Francisco, CA, 1991, EPRI.

4. R.B. Scarlin and P. Schepe: 'State of European COST-activities', *Proc. Conf. Improved coal-fired power plants*, Palo Alto, USA, 1988, EPRI.

5. W. Gysel, A. Trautwein and K.H. Mayer: '10% CrMoWVNbN cast steel for casings for advanced cycles – A collaborate European effort in COST 501–II', *Proc. Conf. Improved coal-fired power plants*, San Francisco, CA, 1991, EPRI.

6. F. Schuster and H. Cerjak: 'Steel castings made from newly developed 9–12% Cr-steels for advanced power generation', *Manuscript Abstract Code No. J33*, ASME Cogen Turbo Power, Vienna, 1995.

7. K.J. Irvine, D.J. Crowe and F.B. Pickering: *The metallurgical evolution of stainless steels*, F.B. Pickering ed., The American Society for Metals, Metals Park, Ohio and The Metal Institute, London. 43–62. (Published originally in *JISI*, 1960, 386–405.)

8. P. Hofer: *Microstructural evaluation of creep processes in a tungsten modified 9–10% chromium cast steel*, Diplomarbeit, TU-Graz, 1994.

9. H. Cerjak, V. Foldyna, P. Hofer and B. Schaffernak: 'Microstructure of advanced high chromium boiler tube steels', *This proceedings volume*.

10. H. Cerjak, P. Hofer and P. Warbichler: 'Microstructural evaluation of aged 9–12% Cr-steels containing W', *Int. Symposium on materials ageing and component life extension*, CISE, Milan, Italy, 1995.

11. H. Cerjak, P. Hofer, P. Thurner and P. Warbichler: 'Quantitative investigation of material M28, M29, M29a, M30 and reference material M10', *Report 8/94*, COST 501/3 WP11–Metallography & Alloy Design Group, TU-Graz, 1994.

12. F. Brühl: 'Verhalten des 9%-Chromstahles X10 CrMoVNb 91 und seiner Schweißverbindungen im Kurz- und Langzeitversuch', *Doctoral Thesis*, TU-Graz, 1989.

13. H. Cerjak and F. Schuster: 'Weldability and behaviour of weldings of new developed creep resistant 9–10% Cr-steels', *Pivisto Italiano dillo Soldtissa 4/94*, 467–473.

Microstructural Development and Stability in New High Strength Steels for Thick Section Applications at up to 620°C

B. NATH,[1] E. METCALFE[1] AND J. HALD[2]

[1] *Engineering, National Power plc, Windmill Hill Business Park, Whitehill Way, Swindon SN5 6PB, UK*

[2] *Department of Metallurgy, Technical University of Denmark, DK–2800 Lyngby, Denmark*

ABSTRACT

An international consortium of steel makers, boiler manufacturers and power producers has developed and validated three new steels which offer almost 50% higher creep rupture strength than P91 at 600°C after 10^5 h. Compositions of these 9–11 Cr steels are based around 1.8–2% W and 0.5% Mo and alloying additions are optimised for the required combination of properties. Two of the three steels have obtained ASME code approval for use as thick section components at up to 620°C.

In the normalised and tempered condition all three martensitic steels exhibit ferrite laths with MC + $M_{23}C_6$ carbides and negligible amounts of δ-ferrite. Intermetallic Laves phase forms during ageing at 600 and 650°C. There is a concomitant decrease in the impact toughness although the tensile properties remain unaffected. By comparison, Laves phase precipitation does not occur in P91 above 600°C. A thermodynamic model has been developed which fully describes the precipitation of the Laves phase.

Creep strengths of NF616 and HCM12A have been compared after two ageing treatments: 650°C for 10 000 h and 720°C for 200 h. The former results in complete precipitation of Laves phase, prior to creep tests. By comparison, Laves phase does not form on ageing at 720°C but it does during long term creep tests at lower temperatures. The 720°C for 200 h specimens exhibit higher rupture strength than those aged at 650°C. The results show that the precipitation of Laves phase during creep is the primary strengthening effect of tungsten in 9–11% Cr steels and that any solid solution strengthening is of secondary importance.

Similar and dissimilar metal welds have been made in both thin and thick section sizes, using different processes e.g. manual metal arc, submerged arc and tungsten inert gas. The heat affected zone (HAZ) exhibits an unusually fine grain size, even adjacent to the fusion line, due to a transformation induced grain refinement. Beyond the inter-critically annealed zone (ICAZ) at the edge of the HAZ, there is a region of minimum hardness which correspond to an over-tempered structure. At moderate to low stresses, creep rupture of cross-weld samples occur at the Type IV location.

123

1. INTRODUCTION

Two main driving forces for increasing the thermal efficiency of fossil fired power stations are the environmental requirements of reduced emissions and the commercial imperative of cheaper power. Significant increases in efficiency are attainable by increasing the steam temperature and/or the pressure to supercritical conditions. Thick section components in such plants would require materials with high creep strength to ensure economic life, high resistance to thermal fatigue for flexible operation, and good fabricability for economic construction of complex components. This has led to considerable effort on developing high strength 9–12Cr steels.

Grade 91 can be used for thick section headers for plant being constructed to operate at temperatures up to 580°C.[1] However, the properties of the material would be inadequate for much higher temperatures and pressures and significantly stronger steels would be required. Therefore, an international consortium of steelmakers, boiler fabricators and electricity producers was established, under EPRI Project RP 1403, to develop and validate steels with approximately 50% higher creep rupture strengths than Grade 91. Accordingly, the consortium set the following target criteria for the new steels (Table 1).

Three steels, NF616, HCM12A and TB12M have been developed and validated for thick section components. Sufficient long term creep rupture data was developed for NF616 and HCM12A for the two materials to gain ASME Code approval.[2] At 600°C, the allowable strengths of the two steels are 30–35% higher than that of the modified-9Cr.[3,4] However, compared against the properties of thick section P91, expected to be around 90 MPa for the

Table 1 Target criteria for the steels

Property	Target
100 000 hour rupture strength at 600°C	>140 MPa
100 000 hour cross weld rupture strength at 600°C (or 30%) strength loss allowed)	>100 MPa
20°C impact strength (unaged)	>40 J
0.2% proof strength at 20°C	>400 MPa
0.2% proof strength at 600°C	>250 MPa
Ultimate tensile strength at 20°C	>600 MPa
Ultimate tensile strength at 600°C	>350 MPa
Elongation	>20%
Reduction in area	>40%
Reduction in area after 10 000 hours at 600°C	>40%
Minimum average tempering temperature (for thick section components a variation of ±20°C will be allowed around this average value)	>760°C

100 000 hour life at 600°C, the strength advantage of NF616 and HCM12A is about 50%.

The detailed scope of the programme and the results have recently been published.[5] The purpose of this paper is to describe the microstructure of parent metals and weldments, and discuss the implications of the microstructural stability on long term creep rupture strengths.

2. MATERIALS

Thick section pipes (350 mm OD × 50 mm wall) have been produced in three 9–11Cr steels (Table 2). All three steels are fully martensitic[3,4,6] although TB12M can have a trace (<3%) of δ-ferrite.[6] A cooling rate in excess of 8°Ch^{-1} is required for a fully martensitic structure in TB12M.

There is a very good correspondence between the equilibrium transformation temperatures, A_{e1} and A_{e3}, calculated by Thermocalc[7] and experimentally measured values of A_{c1} and A_{c3}[3,4,6] (Table 3).

2.1 Microstructure

The steels were normalised at 1050–1100°C and tempered at 770–780°C (Table 3). In this condition, the microstructure revealed tempered martensite (Figs 1–3), which consisted of carbide precipitates in a matrix of small ferrite subgrains with a high density of dislocations. $M_{23}C_6$ carbides formed at prior austenite grain boundaries and subgrain boundaries whereas MX particles, rich in Nb and V, precipitated mainly on dislocation within subgrains. A few small holes, believed to formed by leaching of Cu precipitates during electropolishing, were observed in thin foils of HCM12A. Some needles of AlN have also been observed, proportionately more in HCM12A because of the relatively higher amounts of Al than the other two steels.

Table 2 Nominal compositions of the three steels

Alloy	C	Cr	Mo	W	V	Nb	N	Al	Others
NF616	0.1	9	0.5	1.8	0.2	0.06	0.05	0.007	
HCM12A	0.1	11	0.4	2	0.2	0.05	0.06	0.022	1 Cu
TB12M	0.1	11	0.5	1.8	0.2	0.06	0.06	0.004	1 Ni

Table 3 A comparison of transformation and heat treatment temperatures

Alloy	Calculated		Measured		Normalising temp, °C	Critical cooling rate, °C/h	Tempering temp, °C
	A_{e3}, °C	A_{e1}, °C	A_{c3}, °C	A_{c1}, °C			
NF616	878	821	900	840	1065–1100	29	770–780
HCM12A	832	766	904	805	1050		770
TB12M	875	759	950	785	1080	8	770–775

Fig. 1(a)   Tempered martensite in as received NF616.

Fig. 1(b)   Laves phase and carbides in NF616 after 650°C/10 000 hour ageing.

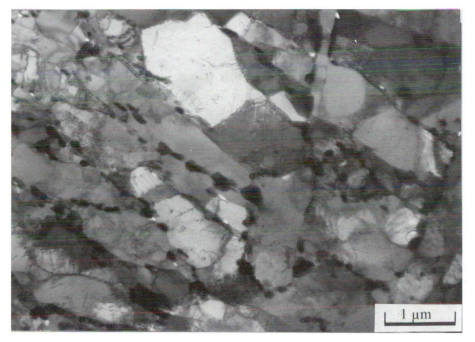

Fig. 2(a) Normalised and tempered HCM12A. Holes are believed to have formed by leaching of Cu precipitates.

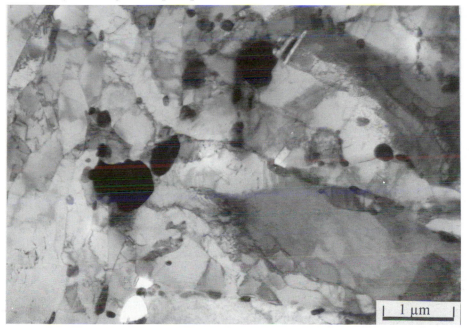

Fig. 2(b) HCM12A after 650°C/10 000 hour ageing showing large particles of the Laves phase.

Fig. 3(a)   As received TB12M.

Fig. 3(b)   Large Laves phase particles in TB12M after 650°C/10 000 hour ageing.

On ageing at 600–650°C for up to 10 000 hours, the dislocation density decreased, the carbides coarsened and the subgrains became a little larger, as expected. However, the most noticeable change in the microstructure was the precipitation of Laves phase of up to 1 μm in size (Figs 1–3). The composition of the particles was consistent with $(Fe,Cr)_2(Mo,W)$. It should be noted that smaller amounts of the Laves phase also formed in P91 but only at 600°C and not at 650°C. The significance of the observation is discussed below.

2.2 Thermodynamic and Kinetics Modelling

Thermocalc calculations have been used to determine the identity and the amounts of different equilibrium phases in the three steels (Figs 4a–4c). Similar calculations have also been performed on P91 for comparison (Fig. 4d). Some of the salient points which emerged are as follows:

- MX is present in all four steels at the normalising temperature in small quantities. This is believed to prevent grain growth during normalising. The proportion of the precipitate increases at lower temperatures to around 0.3% in mass. It follows that only a small proportion of the MX phase would dissolve if the normalising temperature is too low and on subsequent tempering the four materials may not attain the full strength.
- Approximately 2% (by mass) $M_{23}C_6$ precipitate forms in all four steels below A_{e1} and this proportion does not change very significantly below the tempering temperatures.
- The Laves phase precipitates in all three W-containing steels and the amount increases as the temperature falls below around 720°C. Approximately 2% (by mass) precipitates would form at 600°C. Calculations suggest that some of W in $M_{23}C_6$ dissolves to form the Laves phase but the experimental measurements show that the W-content of the carbide does not vary with ageing treatment. This implies that W in the Laves phase comes from the ferrite matrix.

    The Laves phase is predicted in P91 only below 489°C. However, it has been found experimentally at 600°C. The reasons for the discrepancy is believed to be inaccuracy in the Thermocalc database.
- Two other phases, around 0.1% M_2X and 0.5–0.8% Cu, precipitates out in HCM12A below A_{e1}.

The kinetics of precipitation of the Laves phase has been modelled using a Johnson-Mehl-Avrami type relationship. Assumptions for modelling the kinetics were that (i) the bulk diffusion of W in ferrite is the rate controlling process and (ii) the nucleation sites are independent of temperature range under consideration and that all sites are instantly active. Calculations predict sigmoidal kinetics such that after 10 hours ageing approximately 0.25%W (by weight) would partition into carbides irrespective of the ageing temperature.

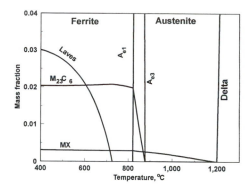

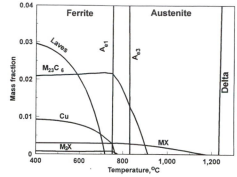

Fig. 4(a) Thermocalc prediction of equilibrium phases in NF616.

Fig. 4(b) Equilibrium phases in HCM12A predicted by Thermocalc.

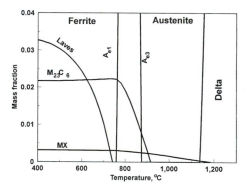

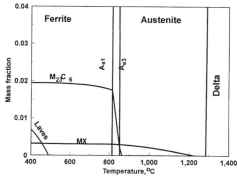

Fig. 4(c) Equilibrium phases in TB12M according to Thermocalc.

Fig. 4(d) Thermocalc prediction of equilibrium phases in P91.

The amount of W in precipitates increases to an equilibrium value which depends on the temperature (Fig. 5). For example, W in precipitates (in both carbides and the Laves phase) reaches an equilibrium value of approximately 1.25% after around 20 000 hours at 600°C, leaving around 0.6% W in solid solution. By comparison, 0.85%W would remain in solution at 650°C. It has already been noted that at equilibrium approximately 2% (by mass) $M_{23}C_6$ forms. Energy dispersive analysis has shown that the precipitate contains around 15% W by weight. The data can be used to estimate approximate partitioning of W in different phases (Table 4). It is clear that around 0.6–0.85%W would remains in solid solution after equilibrium precipitation of carbides and the Laves phase.

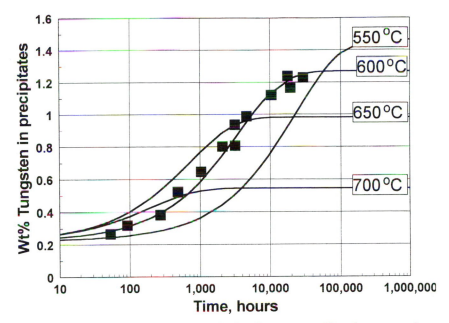

Fig. 5   Sigmoidal kinetics of W-precipitation in NF616. Also shown are the experimentally measured values after ageing at 600°C.

Table 4   Partitioning of W in different phases

Time	W in $M_{23}C_6$	W in Laves		W in solution	
		600°C	650°C	600°C	650°C
10 h	0.25	0	0	1.6	1.6
Equilibrium	0.3	0.95	0.7	0.6	0.85

The calculated and measured partitioning of W in precipitates agree well. The C-curve kinetics also show that the Laves phase does not form on ageing at 720°C (Fig. 6).

2.3 Mechanical Properties

2.3.1 Charpy energy. In the normalised and tempered condition the fracture appearance transition temperature (FATT) of all four steels was either 0°C or below (Fig. 7). In W-containing steels it increased to up to 80°C after up to 10 000 hours ageing at 600–650°C. By comparison, the highest FATT in P91 was 18°C after 600°C for 10 000 h and −18°C after 650°C for 10 000 h ageing. The embrittlement of the W-containing steels is attributed to the precipitation of the Laves phase. This precipitate forms in P91 in smaller quantities and then only at 600°C, not at 650°C. Consequently, the embrittlement of P91 is

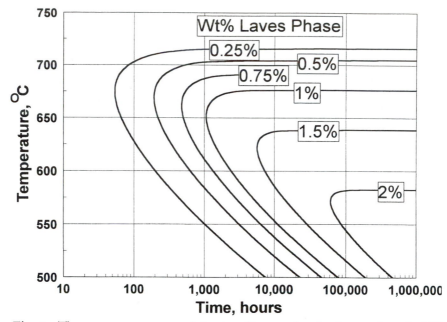

Fig. 6   Time-temperature-precipitation diagram for the Laves phase in NF616.

less marked than in the three W-containing steels. Precipitation of the Laves phase is also responsible for higher embrittlement of P91 at 600°C than at 650°C.

2.3.2 Rupture strength. In order to investigate the microstructural stability and its effects on the rupture properties, iso-stress tests have been conducted on NF616 and HCM12A after three different heat treatments.

- 740°C/2 hours: post weld heat treatment condition, as the reference material it possessed similar rupture strength as the normalised and tempered material.
- 720°C/200 hours: no Laves phase formed and approximately 1.6%W remained in solid solution.
- 650°C/10 000 hours: precipitation of the Laves phase was complete and the solid solution contained only 0.6%W.

The Larson-Miller parameters of the 720°C/200 hours and 650°C/10 000 hours treatments were identical. Consequently the steels exhibited similar hardness and the tensile strengths after the two heat treatments. Rupture lives of the steels at 660°C were identical after the two ageing treatments (Fig. 8). Test durations of up to 100 hours were too short for any significant change in precipitates. Thus, the material with 1.6%W in solution had the same strength as the steel with only 0.6%W i.e. the solid solution strengthening effect of W

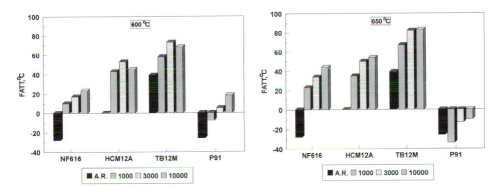

Fig. 7   FATT after different heat treatments.

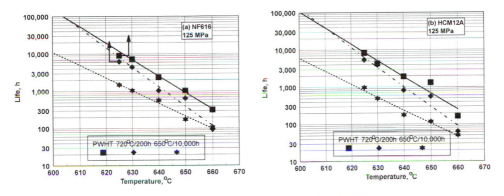

Fig. 8   Iso-stress rupture data for (a) NF616 and (b) HCM12A.

is relatively small or that 0.6 weight percent is sufficient to realise the full potential. By a process of elimination, it is concluded that the primary strengthening role of W is the precipitation of the Laves phase.

It should be noted that 0.6 and 0.85% W would remain in solid solution after the materials have reached equilibrium at 600 and 650°C respectively.

The rupture strength after 650°C/10 000 hours, equivalent to 600°C/200 000 hours service, was similar to P91 in the virgin condition. This indicates that the creep rupture strength of the three W-containing steels does not decrease catastrophically due to the precipitation of the Laves phase and that the extrapolated strengths can be regarded with confidence.

2.4 Weldments

2.4.1 Fabrication. The three steels exhibit very good weldability. Procedures have been developed for similar and dissimilar metal welds in these steels using processes ranging from tungsten inert gas (TIG, also known as gas

Table 5 Similar and dissimilar metal welds in NF616, HCM12A and TB12M[3,8,9,10]

Parent 1	Filler	Parent 2	OD	Wall	Process	Min. preheat	Max. interpass	PWHT temp, °C
NF616	NF616	NF616	350	50	MMA, SAW	200	350	740
HCM12A		HCM12A	380				300	750
TB12M		TB12M	350			200		
NF616	NF616	P91	350	50	MMA	150	300	750
HCM12A								
TB12M			380					
NF616	Grade 91	P91	350	50	MMA + SAW	200	350	740
HCM12A								
NF616	Inco 182	NF709	54	12	MMA	150	300	750
HCM12A		HR3C						
TB12M		Eshette 1250						
NF616	Inco 82	NF709	54	12	TIG	150		715
HCM12A		HR3C						
NF616	Type 309	NF709	54	12	TIG	150		715
HCM12A		HR3C						

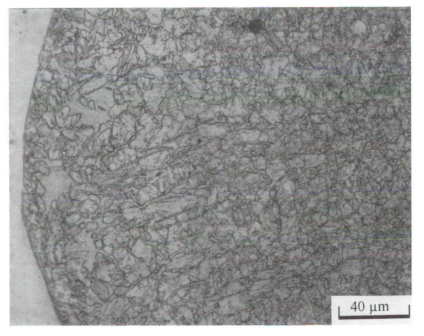

Fig. 9(a) Optically sharp *Type I* interface dominates ferritic-Inconel 182 boundary. Very fine grained HAZ and δ-ferrite are also evident.

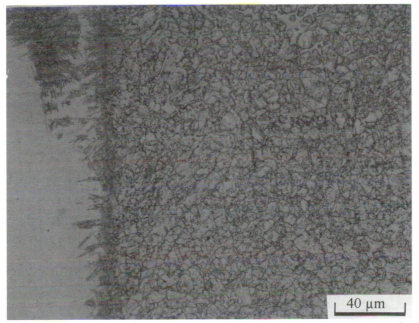

Fig. 9(b) The tempered martensitic structure of the *Type II* interface in ferritic-austenitic joints.

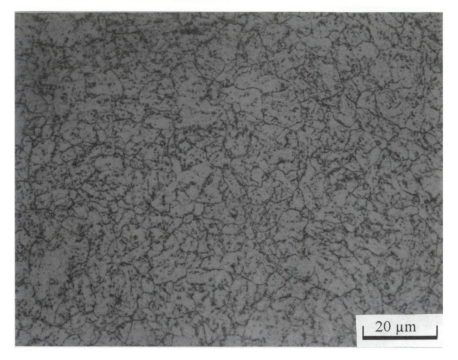

Fig. 10 A typical structure of the ICAZ.

tungsten arc or GTAW), manual metal arc (MMA or SMAW) and submerged arc (SAW)[3,4,8,9,10] (Table 5). The early development work on similar metal welds in the three 9–11 Cr steels were conducted on joints made with the NF616 filler but matching consumable is now available for HCM12A.

Use of these high strength steel as headers would involve dissimilar metal welds between the W-containing steels and austenitic stainless steel tubes. In some instances the steels may also have to welded to thick section components in P91. Either an austenitic (Type 309) or a nickel-based filler (Inco 82 or Inco 182) were used to weld the ferritic-austenitic tubes whereas pipe size joints with P91 were made with either NF616 or a Grade 91 consumable.[10]

In general, preheats in excess of 150°C and interpass temperatures of 300–350°C have been used. The welds are stress relieved at 740–750°C. Stress relieving treatment is considered mandatory for these steels, even in tube sizes, because on bending, as-welded tube joints cracked at the fusion line.[10] In bend tests, all stress relieved welds satisfied the ASME Section IX acceptance criteria of a 180° angle bend.

2.4.2 Microstructural characterisation. The fusion boundary between the W-containing steels and the nickel-based filler metal is typical of ferritic-austenitic joints of this type. Most of the interface is optically sharp and is

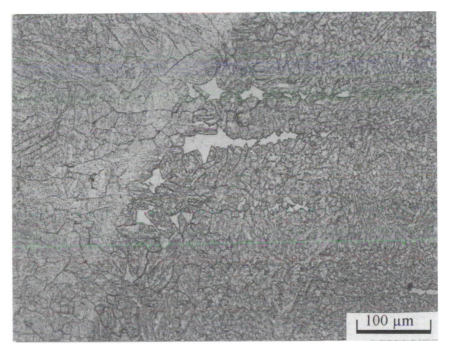

Fig. 11 The interface between P91 and the NF616 weld metal.

known as the Type I interface (Fig. 9a). The remaining Type II interface is a diffuse band of tempered martensite (Fig. 9b).

The HAZ in the three W-containing steels were unusually fine grained, even in the immediate vicinity of the fusion boundary (Fig. 9). The HAZ was predominantly martensitic but close to the fusion line, thin films of δ-ferrite were present at martensite grain boundaries. Some equiaxed grains of δ-ferrite also formed near the fusion boundary and often these exhibited multiple islands of martensite (Fig. 9). This suggests that during the welding cycle, the material close to the weld bead transforms to δ-ferrite. On cooling, multiple grains of austenite nucleate in the δ-phase and the subsequent transformation of the austenite produces intragranular martensite islands. Individual grains of austenite can grow to consume most of the parent δ-grain leaving a thin film of the δ-phase at grain boundaries. It is suggested that such a sequence of transformation during the welding cycle refines the grain size of the HAZ.

The ICAZ was typically at 3–3.5 mm from the fusion boundary. The microstructure was characterised by small grains decorating prior austenite grain boundaries, like beads in a necklace (Fig. 10). The hardness of the ICAZ was approximately 230–240 VPN.

Both the P91 HAZ and the NF616 filler were mainly martensitic, with a few δ-ferrite grains near the fusion boundary (Fig. 11). The continuation of

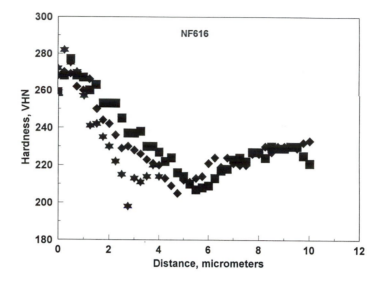

Fig. 12(a) NF616/Inco 182/NF709 joint: hardness profile across NF616 HAZ at three locations.

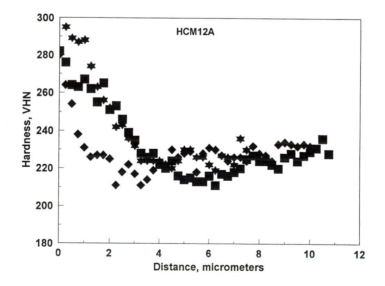

Fig. 12(b) HCM12A/Inco 182/HR3C weld: hardness variations in HCM12A HAZ at three positions in the wall.

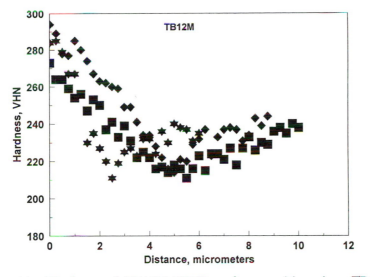

Fig. 12(c)   Hardness of TB12M HAZ at three positions in a TB12M/Inco 182/Eshette 1250 weld.

some grain boundaries across the fusion line indicates a degree of epitaxial growth.

The HAZs in all welds exhibit hardness minima with a typical value of 200–220 VPN. The minima were located at 5–6 mm in tube-size joints (Fig. 12) and at 4–5 mm in pipes (Fig. 13), which has a significantly higher thermal mass. This softening is attributed to overtempering.

2.4.3 Creep failures of cross-weld samples. The failure location of uniaxial cross-weld creep samples appears to be a competitive process such that at a given stress and temperature, the weakest (and/or the most brittle) link in the chain ruptures before others do. Which link is the weakest or the most brittle depends on the stress and temperature as described below for ferritic-austenitic joints.

- Short term failures in the parent metal, remote from the weld, at high stresses and relatively low temperatures.[3,10] Such rupture is typical of cross-weld samples. Failure strains in these samples are relatively high and the samples exhibit necking (Fig. 14).
- The second mode of fracture at high stresses and relatively low temperatures occurred in the weld metal, in NF616-austenitic and some TB12M-austenitic transition joints[10] (Fig. 15). The reasons for this unusual fracture in dissimilar metal welds are not clear but it is possible that the rupture strength or the ductility of the diluted weld pool renders the material prone to creep rupture.

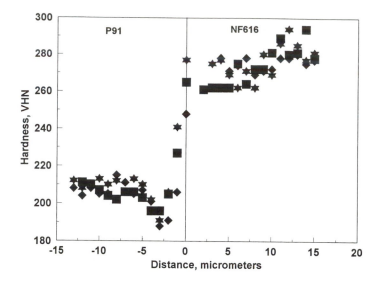

Fig. 13   Dissimilar metal welds between P91 and the W-containing steels: three hardness traverses across the P91 fusion boundary with NF616 filler.

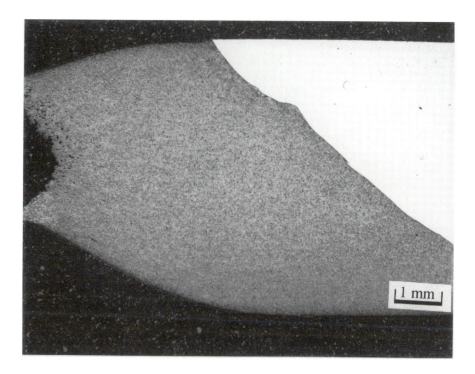

Fig. 14   An example of parent metal failure.

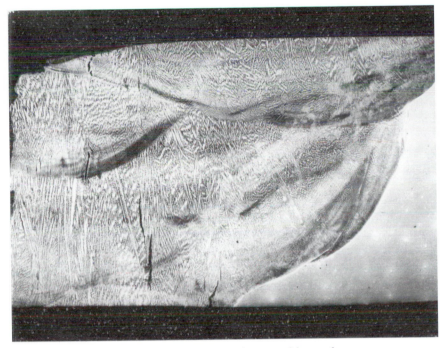

Fig. 15   Rupture and cracks in the Inconel 182 weld metal.

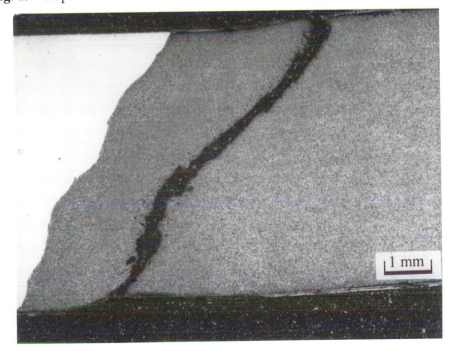

Fig. 16   Type IV fracture at low to moderate stresses.

- Relatively long term failures at stresses and temperatures approaching service conditions occurred at the Type IV location, in the vicinity of the ICAZ[3,10] (Fig. 16). This type of fracture is typical of most low and high alloy ferritic steels both in laboratory tests as well as in service. It should be noted that the Type IV failure does not occur where the hardness in lowest.

Transition joints between W-containing steels and P91 fail in the latter. Short term rupture occurs in the parent P91 and but the ICAZ fails in long term tests.

In common with most materials, the welded joints in the 9–11 Cr steels also exhibit lower strengths than the corresponding parent metal. In general, the rupture curves for the parent metal and the weldments diverge with increasing duration. Thus, the strength reduction factor depends not only on the material but also on the temperature and the duration. For the three steels, the extrapolated values of 600°C/100 000 hour rupture strengths of the welds are up to 20% lower than those of the parent metal.

4. SUMMARY

1. Three W-containing steels, NF616, HCM12A and TB12M, have been developed for use as thick section components. All steels have attained the mechanical property targets set at the beginning of the project and they all exhibit good fabricability.

2. Two steels, NF616 and HCM12A have sufficient long term creep rupture data to have gained ASME Code approval for applications up to 620°C. The 600°C/100 000 hour extrapolated rupture strengths of thick section pipes in these W-containing steels are almost 50% higher than that of P91.

3. The results of thermodynamic and kinetics models agree well with the observed precipitation behaviours. The most significant microstructural change which occurs on prolonged ageing at 600 to 650°C is the precipitation of the Laves phase. W in the Laves phase comes from the solid solution. Approximately 0.6% and 0.85% W would remain in solid solution even after the materials attain equilibrium on prolonged ageing at 600 and 650°C respectively.

4. The solid solution strengthening effect of W is small. The precipitation of the Laves phase during creep deformation does not lead to a catastrophic drop in rupture strength of the steels. On the contrary, it is the main mechanism of strengthening in these W-containing materials.

5. All three steels are readily weldable by a variety of commonly used processes and different filler metals. Some δ-ferrite forms in the HAZ, close to the fusion boundary, either as equiaxed grains or as thin films at prior austenite grain boundaries. The grain size in this region is unusually fine because of transformation-induced grain refinement.

6. In short term creep tests at relatively high stresses and temperatures cross-weld samples fail in either the parent or the weld metal. In longer term, Type IV rupture occurs in the vicinity of the ICAZ. The 600°C/100 000 hour rupture strength is up to 20% less than the parent metal.

ACKNOWLEDGEMENT
This paper is published by permission of National Power Plc and ELSAM and is based on the research carried out by the partners in EPRI RP1403–50 project.

REFERENCES
1. E. Metcalfe and R. Blum: *Proc. Conf. New steels for advanced plant up to 620°C*, London, 11 May 1995. EPRI/National Power.

2. F. Masuyama: *Proc. Conf. New steels for advanced plant up to 620°C*, London, 11 May 1995. EPRI/National Power.

3. H. Naoi, H. Mimura, M. Ohgami, H. Morimoto, T. Tanaka and Y. Yazaki: *Proc. Conf. New steels for advanced plant up to 620°C*, London, 11 May 1995. EPRI/National Power.

4. Y. Sawaragi, A. Iseda, K. Ogawa, F. Masuyama and T. Yokoyama: *Proc. Conf. New steels for advanced plant up to 620°C*, London, 11 May 1995. EPRI/National Power.

5. E. Metcalfe and W.T. Bakker: *Proc. Conf. New steels for advanced plant up to 620°C*, London, 11 May 1995. EPRI/National Power.

6. G.A. Honeyman: *Proc. Conf. New steels for advanced plant up to 620°C*, London, 11 May 1995. EPRI/National Power.

7. J. Hald: *Proc. Conf. New steels for advanced plant up to 620°C*, London, 11 May 1995. EPRI/National Power.

8. F. Masuyama and T. Yokoyama: *Proc. Conf. New steels for advanced plant up to 620°C*, London, 11 May 1995. EPRI/National Power.

9. P.J. Bygate and P.M. Reynolds: *Proc. Conf. New steels for advanced plant up to 620°C*, London, 11 May 1995. EPRI/National Power.

10. B. Nath and F. Masuyama: *Proc. Conf. New steels for advanced plant up to 620°C*, London, 11 May 1995. EPRI/National Power.

Microstructure of Advanced High Chromium Power Plant Steels

H. CERJAK,* V. FOLDYNA,** P. HOFER,*
B. SCHAFFERNAK*
* Technical University Graz, Austria
** Vitkovice j.r.s. Research Institute, Czech Republic

ABSTRACT

Mainly for temperatures above 560°C there is a great demand to replace the widely used austenitic tube materials through advanced ferritic-martensitic steels, in order to utilise their better thermal conductivity, lower thermal expansion coefficient, lower costs and better resistance against stress corrosion cracking susceptibility. An overview of the available steels and their mechanical properties is given. The development of the microstructure at service conditions was investigated qualitatively and quantitatively by hardness tests, light and electron microscopic examination, including EDX investigation. The microstructural development predicted by the thermodynamical program package *Thermo Calc* was compared with the experimental results. The effect of the microstructural changes, particularly the heavy precipitation of Laves phase, on the strengthening mechanisms is described. Common used methods for life time prediction are evaluated critically taking into account the changes of mechanisms. A strategy for further modelling activities is given in a flowchart.

1. INTRODUCTION

On the one hand, the demands for higher steam parameters to increase the efficiency of thermal power plants also affect the requirements for boiler tube materials. Mainly for temperatures above 560°C there is a great demand to replace the widely used austenitic tube materials through advanced ferritic-martensitic steels, in order to utilise their better thermal conductivity, lower thermal expansion coefficient, lower costs and better resistance against stress corrosion cracking susceptibility. On the other hand, the user is faced with typical boiler tube problems concerning fabrication and service. In fabrication, the HAZ hardening during welding, the related requirements for applying a PWHT as well as the capability for a bending procedure are of interest. For the service, the oxidation in the fireside and the flue gas atmosphere become relevant.

A number of advanced ferritic-martensitic steels have appeared in the last decade which show promising behaviour, mainly regarding their improved creep strength compared to the materials used until now, 2.25 Cr 1 Mo and

Table 1 Chemical composition of 9–12% Cr-steels

	X20CrMoV 12 1	P/T 91	NF616 P/T 92	HCM 12	HCM 12A P/T 122	E 911	COST Steel D3
C	0.17–0.23	0.08–0.12	0.06–0.13	max. 0.14	0.06–0.14	0.09–0.13	0.16
Mn	max. 1	0.30–0.60	0.30–0.60	0.30–0.70	max. 0.70	0.30–0.60	0.49
P	0.030	max. 0.02	max. 0.02	max. 0.30	max. 0.03	max. 0.020	
S	0.030	max. 0.01	max. 0.01	max. 0.30	max. 0.02	max. 0.010	
Si	max. 0.5	0.20–0.50	max. 0.5	max. 0.50	max. 0.70		0.12
Cr	10.0–12.5	8.00–9.50	8.00–9.50	11.0–13.0	10.00–12.60	8.50–9.50	11.3
Mo	0.80–1.20	0.85–1.05	0.30–0.60	0.80–1.20	0.20–0.60	0.90–1.10	0.32
W			1.50–2.20	0.80–1.20	1.50–2.50	0.90–1.10	1.8
Ni	0.30–0.80	max. 0.40	max. 0.40		max. 0.70	0.10–0.40	0.79
V	0.25–0.35	0.18–0.25	0.15–0.25	0.20–0.30	0.15–0.30	0.15–0.25	0.22
Nb		0.06–0.10	0.03–0.10	max. 0.20	0.02–0.10	0.06–0.10	0.06
N		0.03–0.07	0.03–0.09		0.02–0.10	0.06–0.08	0.06
Al		max. 0.04	max. 0.04		max. 0.04	max. 0.025	
B			max. 0.006		max. 0.005		
Cu					0.48–1.56		

X20 CrMoV 12 1. The advanced 9 to 12% Cr-steels show superior creep resistance compared to the reference material X20, which has been used very successfully over the past three decades. The creep resistance of the American development grade P/T 91 is surpassed by the W-containing grades NF616 and HCM 12A, recently approved by ASME: CC 2179 as P/T 92 and CC 2180 as P/T 122.[1] The European version E911 is under investigation in Round III of the COST Action 501.[2] In addition, results on W-containing forgings and castings are reported extensively from COST 501, Round II.[3]

Table 1 gives an overview of the chemical compositions of conventional and advanced ferritic chromium steels, Fig. 1 shows the allowable stresses for some of this steels.

2. MICROSTRUCTURE AND STRENGTHENING MECHANISMS

The mechanical properties as well as creep rupture strengths of Cr-modified steels result from a balance of their chemical composition and their microstructure. It is evident that not only the chemical composition but also the heat treatment influences the creep properties very significantly. In addition, microstructural changes occur during service of these materials in the applied temperature and stress ranges, depending on the chemical composition and the initial microstructure.

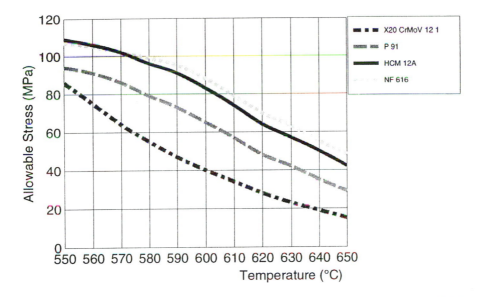

Fig. 1   Comparison of maximum allowable stress for 9–12% Cr-steels (ASME, DIN).

The strengthening mechanisms acting in these steels can generally be described as a combination of solid solution strengthening, dislocation–particle interactions, dislocation–dislocation and dislocation–subgrain boundary interactions.

From this, it can be seen that there is a complex relationship between the microstructural development during the life time of the component and the development of the strengthening mechanism acting, dependent on the stress level applied. To predict the long term behaviour of this materials, it is necessary to understand the principal correlations between microstructure and strengthening mechanisms as well as the microstructural changes and the effect of these on the strengthening mechanisms. In addition, at the end of the lifetime an embrittlement of the microstructure can reduce the allowable stress because of low ductility fracture behaviour. Only by understanding this relationship can a controlled development with promising results be expected. These implications have to be understood not only for the base materials but also for the heat affected zone of weldings and the deposited weld material.

A great deal of information is available in the literature about the microstructures in the as received as well as in the long-term exposed condition of different 9–12% Cr-steels.[2,4] In Table 2 a very rough qualitative description of this observation is given. The as received microstructure of the modified 9 to 12% Cr-steels can be described as tempered martensite with high dislocation

Table 2 Qualitative description of the microstructural development of W and Mo containing 9–12% Cr-steels under service conditions, 600–650°C.

		As received	Exposed >10 000–30 000 h	Strained >10 000–30 000 h
Hardness		high	≈95% of as received	≈70–80% of as received
Light microscopy		tempered martensite, $M_{23}C_6$, δ-ferrite depending on the grade	tempered martensite, $M_{23}C_6$, partly decomposed δ-ferrite	tempered martensite, $M_{23}C_6$, partly decomposed δ-ferrite
TEM	Dislocation density	high	low	very low
	Subgrain/ martensite lath size width	small lath width	martensite laths transform to subgrains	subgrains fully recovered
Particles:	$M_{23}C_6$	on lath boundaries typically 50–150 nm	partly coarsened on subgrains boundaries	partly coarsened on subgrain boundaries
	MX	fine dispersed (≈20–50 nm)	fine dispersed (≈20–50 nm)	fine dispersed (≈20–50 nm)
	Laves phase	no Laves phase	Laves Phase medium and large size precipitates (≈200–500 nm)	Laves phase large size precipitates (≈500 nm)

density, fine subgrain sizes and $M_{23}C_6$ precipitations. In some grades, delta-ferrite up to 30% can be present. In the case of V and Nb containing grades, very fine dispersed MX precipitates of vanadium-nitrides and/or Nb carbonitrides appear inside the subgrains and interact with dislocations.[4,5,6] Extensive round-robin metallographic investigations performed on these types of materials by the Metallography and Alloy Design Group of COST 501, Round III, which is still in progress[2] revealed that the following effects could be observed depending on time, temperature and stress/strain during exposure in the creep range:

- Recovery of the martensitic structure by lowering the dislocation density and coarsening of the subgrains,
- Precipitation of W and Mo containing Laves phases and coarsening of this precipitation.
- Depletion of the matrix from W and Mo.[7]
- A coarsening of $M_{23}C_6$ particles but not as pronounced as observed at the Laves phases.

The grade of the change in the microstructure is dependent on the temperature of exposure, the time of exposure, the strain/stress level applied[7,8] and the thermodynamic driving forces controlling the kinetics of the evolution of the microstructure. Because of the similar alloying principles of this steel grades

Table 3 Comparison of used procedures for assessing the creep rupture strength of the steel 0.13%, 8.9% Cr and 2.16% Mo[15]

External conditions	Stress dependence $P_{LM} = f(\log \sigma)$	Constant C_{LM}	Number of results	Creep rupture strength in 10^5 h at 600°C
Low stress domain $\sigma \leq \sigma_z(T)$	linear	22	40	47
High stress domain $\sigma \leq \sigma_z(T)$	linear	31	22	78
Throughout the temperature and stress interval	polynomial	23.9	58	46.9

the thermodynamic driving forces must be similar. The different chemical compositions and heat treatment conditions will be of influence both on the kinetics of the microstructural development and the thermodynamic driving forces. Therefore in this paper no distinction will be made regarding the product form (boiler tube, forging, cast) of similar types of chemical composition.

A widely used method to predict the longtime creep rupture behaviour is the application of Larson-Miller parameters.

As the derivation of Larson-Miller parametric equation is based on the Arrhenius equation, the constant C_{LM} has its physical meaning and represents the pre-exponential factor in the Arrhenius equation.[9] It is therefore clear that when changing the principal deformation mechanism the respective C_{LM} constant has to be changed. With increasing temperature and decreasing stress, the activation energy of deformation or fracture decreases as well as the pre-exponential factor. Therefore, one can expect lower C_{LM} too.

The well known Ashby creep deformation or creep rupture maps can be taken for determining the temperature-stress domain in which the only one deformation mechanism acts and in which the extrapolation by using C_{LM} constant is possible. When crossing the domain boundary another deformation mechanism prevails with different activation energy and stress exponent and therefore extrapolation with previously used C_{LM} is incorrect and leads to significant overestimation of creep rupture strength. An evaluation of this observations is given in Table 3.[15] Recently have been published relatively long term creep rupture tests attained on the steel D3.[10] These results are convenient to show how significant overestimation can be expected using different values of $C_{LM} = 25$ and $C_{LM} = 35$. The constant $C_{LM} = 25$ is used in the current Round of COST 501, $C_{LM} = 35$ is also often used and the following parameter $P = T^* (35 + \log t)$ is sometimes called as Masuyama parameter.[11] Comparing experimentally attained and predicted times to rupture, there is

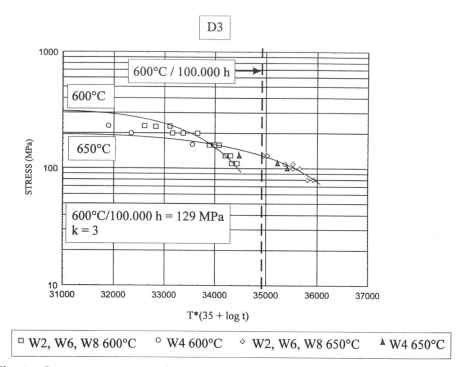

Fig. 2   Creep rupture strength extrapolation using Masuyama parameter[10] WE ... W8: Different heat treatment conditions of COST steel D3.

evidence that both used values of C_{LM} significantly overestimate the predicted time to rupture (Table 4). Relation between stress and Masuyama parameter is shown in Fig. 2. K = 3 means the highest exponent used in the stress dependent part of the Larson-Miller equation:

$$P_{LM} = a_0 + a_1 \cdot \log(\sigma) + a_2 \cdot \log(\sigma)^2 + a_3 \cdot \log(\sigma)^3 = T \cdot (C_{LM} + \log(t))$$

It is clear, that based on creep rupture tests performed at 650°C the predicted time to rupture at 600°C is about 15 times longer than the experimentally established. Good agreement between experimentally ascertained and predicted creep rupture times can be attained when C_{LM} is about 14. The most probable estimation of creep rupture strength estimation of mentioned steel grade can be expected using C_{LM} = 14.4 (Table. 5). Nevertheless, for more reliable creep rupture strength estimation can be recommended to perform much longer creep rupture tests and at more than two temperatures.

3. MODELLING OF THE MICROSTRUCTURAL DEVELOPMENT
To overcome the problem of fulfilling the demand of designers, to offer reliable long term creep properties for newly developed steels in short amount of time and additionally lower the risk of producing non-reliable figures as well as a

Table 4 How the predicted time to rupture at 600°C and 110 MPa depends on the C_{LM} constant; steel D3 – COST 501

Stress	Temperature	Attained time to Rupture*	Predicted time to rupture [h]		
[MPa]	[°C]	[h]	$C_{LM} = 25$	$C_{LM} = 35$	$C_{LM} = 14$
110	600	23 000	94 300	355 550	22 250
	650	2 260			

* Mean values

guide for the development of a promising alloy concept, the modelling of the microstructures based on thermodynamic equilibrium calculations can be used.

The equilibria in multicomponent multiphase systems can be calculated by minimising the Gibbs free energy of the system using thermodynamic data extrapolated from a database containing thermodynamic functions for alloy subsystems. The program package *Thermo Calc*[12] using the SGTE data base[13] is one of the readily available tools for metallurgists for calculating thermodynamic equilibria and also the driving forces for phase transformations (the reduction of the Gibbs free energy during the transformation).

The calculation of phase diagrams and the prediction of the chemical composition of phases are very helpful for the investigation of the microstructure of materials, because they can show us which direction the evolution of the microstructure will go. Unfortunately it is not possible to figure directly the way and the kinetics for the transformations. To obtain this information, we need kinetic models for nucleation, growth and ripening. The equilibrium composition and the driving forces calculated by *Thermo Calc* are important input parameters for kinetic models.[14]

Performing parameter studies using *Thermo Calc* one obtains very accurate results showing the dependence of the phase diagrams and phase compositions from the changes of one or more parameters (e.g. the change of the amount of one alloying element). But the absolute values obtained by the equilibrium calculations can be strongly different from the results found experimentally. There are several reasons for this observation:

- The transformation temperatures calculated by *Thermo Calc* are equilibrium transformation temperatures, the effect of undercooling and superheating is not taken into account.
- The results are dependent on the assessment of the thermodynamical functions stored in the database and on the models used for the interpolation of the thermodynamic data.[13]
- The database used is not complete for all elements appearing in ferritic chromium steels.

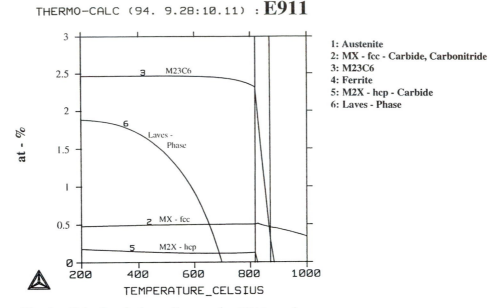

Fig. 3 Calculated phase diagram for E911 steel.

Calculations and predictions made using a thermodynamical computer program must of course be verified by experiments, but an experimental procedure with trial and error can be avoided. Parameter studies are not only valid for one alloy composition, but for all compositions with the same basic mechanisms.

Figure 3 shows an example for a calculated phase diagram (E911). The results are expressed in terms of at-% or mole fractions and can be directly used in further calculations. Since the database is not complete for all elements, the input for *Thermo Calc* can be simplified: P, S and B are not considered, for the different MX precipitates a general model for a fcc phase is used, which cannot distinguish the different types of MX precipitates.

The results can be phase diagrams or tables showing the composition of the different phases in the equilibrium state.

4. COMPARISON OF MODELLING AND EXPERIMENTAL RESULTS: MATRIX DEPLETION

As the elements tungsten and molybdenum are bound mainly in the Laves phase, simultaneously with the formation of the Laves phase, a matrix depletion of these alloying elements can be calculated for the discussed chromium steels and compared with experimental results. Experimental results are shown in Fig. 4 for the 1% W and 1% Mo containing cast steel G–X12CrMoWVNbN10–1–1 after exposing at different temperatures and

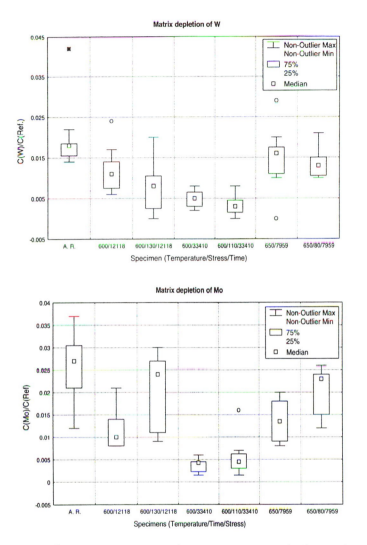

Fig. 4 Material GX–12CrMoWVNbN10–1–1: Matrix depletion during ageing and creep.[8]

times up to 33 410 hours. Variations of matrix composition owing to the carbide and Laves phase formation were examined by quantitative EDX analysis. The results obtained are summarised in Box Plots to represent the change in the Mo and W content in the ferritic matrix, because these elements contribute extensively to the coarsening of $M_{23}C_6$ carbides and formation of Laves phase. A more detailed description of these investigations is given in reference.[7,8] Note: The Y-axis in Fig. 4 is inscribed with the ratio 'intensity of Mo(W) peak/intensity of Fe peak multiplied by the k factor'.

These results are in very good agreement with the equilibrium calculations.

Table 5 How the predicted creep rupture strength at 600°C in 100 000 h depends on the C_{LM} constant; steel D3 – COST 501

Predicted creep rupture strength at 600°C in 100 000 h [MPa]			
$C_{LM} = 25$	$C_{LM} = 35$	$C_{LM} = 14.4$	$C_{LM} = 19$
99[1]	114[1]		
106[2]	126[2]	74[4]	185[1]
106[2]	129[3]		

$C_{LM} = 25$ and $C_{LM} = 35$ elected constants
$C_{LM} = 14.4$ and $C_{LM} = 19$ calculated constants
[1] all results are taken into account
[2] long term creep rupture tests at 600°C ($\sigma \leq 129$ MPa, $t_r \geq 14\,466$ h) are not taken into account
[3] creep rupture test at 650°C are taken into account, only
[4] creep rupture tests attained in low stress domain (600°C, $\sigma \leq 160$ MPa; 650°C, $\sigma \leq 110$ MPa) are taken into account only

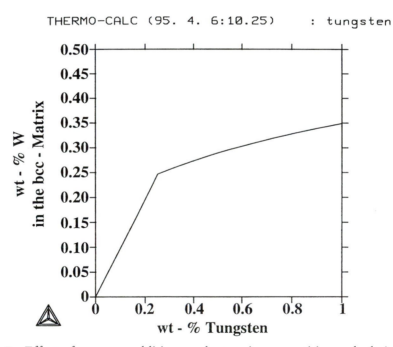

Fig. 5 Effect of tungsten addition on the matrix composition, calculation at 600°C.

The equilibrium amount of tungsten in the bcc-matrix at 600°C predicted by *Thermo Calc* is 0.3513 wt%. This is about 37% of the relative tungsten content in the alloy. The predicted Mo-content in the matrix is 0.3568 wt% (overall composition: 0.97 wt% Mo). Figure 5 shows the effect of tungsten addition on the matrix composition. One can see in this diagram, that the higher part of tungsten is not present in the matrix. So this part cannot contribute to the solid solution strengthening mechanism.

5. FUTURE ASPECTS

Newly developed chromium modified steels are intentionally alloyed with nitrogen. Extensive precipitation of VN and Nb(CN) then augments precipitation strengthening by $M_{23}C_6$ and gives the steel about 40% higher creep rupture strength compared to X20CrMoV12 1. Particles of Nb(C,N) which stay undissolved during austenising pin grain boundaries and restrict austenite grain growth during normalising. To obtain good creep strength it is necessary to keep high nitrogen content in the solid solution and a great attention should be paid to the chemical composition with respect to V, Nb and Al.[17]

In order to verify models for the precipitation, growth and ripening of particles it is necessary to know the history of each kind of particles appearing. A new experimental method, the Energy Filtering Transmission Electron Microscopy (EFTEM)[16] is available now. Some essential advantages become accessible with this method. The contrast and resolution of TEM images and electron diffractions can be improved essentially, precipitates which are hardly recognisable on TEM images can be found and identified with EFTEM. Figure 6 shows an example for an elemental map of NF616 steel created with the EFTEM-method. The Fe distribution can be seen as an extraction of every particle existing in the microstructure, independent of its orientation. It should be noted that some particles visible in the Fe distribution don't appear in the TEM bright field image. Especially particles of the type MX containing N and V are very difficult to find without this new method in the TEM bright field image.

Using this method for several specimens aged for different times one can draw the history of each kind of precipitates (amount, composition, growth, ripening), calculated results can be verified seriously.

Figure 7 proposes a flowchart illustrating the possible sequence of actions for the development of new steels or modifications of available steels in order to obtain optimal properties on the basis of an understanding of the interactions between the as received microstructure, microstructural development, and the related strengthening mechanisms.

One can see that a close cooperation between metallography and modelling is very important. Physical models must describe the microstructural mechanisms influencing creep and creep rupture and predict the evolution of the

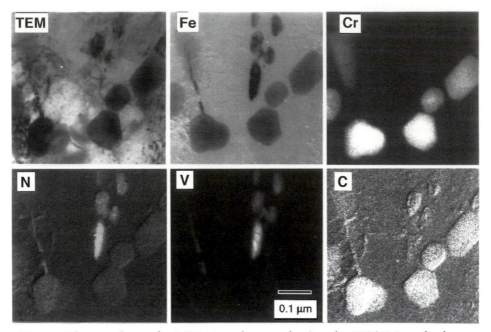

Fig. 6   Elemental map for NF616 steel, created using the EFTEM method.

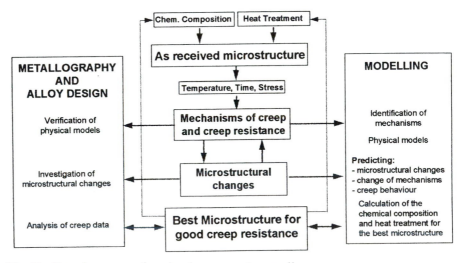

Fig. 7   Development of optimal creep resistant alloys.

microstructure with the goal of finding the best microstructure for an excellent creep resistance. The models are based on the metallographic investigations and must be verified by metallographic methods.

The parameters which can be changed are the chemical composition and the heat treatment of the material: the creep behaviour is influenced by temperature, time and stress.

By applying this procedure, accompanied by experimental verification and extrapolation, more reliable results will be obtained.

ACKNOWLEDGEMENTS

The authors are grateful for the financial support for the research project by the FWF, the 'Fond zur Förderung der wissenschaftlichen Forschung', which has sponsored the work at the Technical University Graz.

The authors wish also to express their thanks for the good cooperation provided by the 'Research Institute for Electron Microscopy' at the Technical University Graz.

REFERENCES

1. F. Masuyama: *Proc. EPRI/National Power Conf. New Steels for Advanced Plant up to 620°C*, London, UK, 1995, Palo Alto, 1995, Paper 8.
2. R.W. Vanstone, H. Cerjak, V. Foldyna, J. Hald and K. Spiradek: *Microstructural Development in Advanced 9–12% Cr Creep Resisting Steels*, A Collaborative Investigation in COST 501/3 WP11, Cambridge, 30 June 1995.
3. *Proc. Conf. Materials for Advanced Power Engineering*, Liège B 1994, D. Coutsouradis *et al.* eds, Kluwer Academic Publishers, Dordrecht/Boston/London, 1994.
4. V. Foldyna, A. Jakobova, V. Vodarek and Z. Kubon: *Proc. Conf. Materials for Advanced Power Engineering*, Liège B 1994, D. Coutsouradis *et al.* eds, Kluwer Academic Publishers, Dordrecht/Boston/London, 1994, Part I, 453–465.
5. F. Brühl, H. Cerjak, P. Schwaab and H. Weber: *Steel Research 62*, 1991, **2**, 75–82.
6. K. Spiradek, G. Zeiler and R. Bauer: *Proc. Conf. Materials for Advanced Power Engineering*, Liège B, 1994, D. Coutsouradis *et al.* eds, Kluwer Academic Publishers, Dordrecht/Boston/London, 1994, Part I, 251–262.
7. H. Cerjak, P. Hofer and P. Warbichler: 'Microstructural evaluation of aged 9–12% Cr Steels containing W', *Materials Ageing and Component Life Extension*, Conf. Milano 10–13 October 1995.
8. K.H. Mayer, H. Cerjak, P. Hofer, E. Letofsky and F. Schuster: *Evolution of Microstructure and Properties of 10% Cr Steel Castings*, Cambridge, 30 June 1995.

9. E. Pink: *Materials Science and Technology*, 1994, **10**, 340.

10. C. Berger, R.B. Scarlin, K.H. Mayer, D.V. Thornton and S.M. Beech: *Proc. Conf. Materials for Advanced Power Engineering*, Liège B, 1994, D. Coutsouradis *et al.* eds, Kluwer Academic Publishers, Dordrecht/Boston/London, 1994, Part I, 47.

11. G.A. Honeyman: *Proc. EPRI/National Power Conf. New Steels for Advanced Plant up to 620°C*, London, UK, 1995, Palo Alto, 1995, Paper 6.

12. B. Sundman: *Anales de física*, 1990, **36B**, 69–81.

13. B. Sundman, B. Jansson and J.-O. Andersson: *Calphad*, 1985, **9**(2), 153–190.

14. J. Hald: *Proc. EPRI/National Power Conf. New Steels for Advanced Plant up to 620°C*, London, UK, 1995, Palo Alto, 1995, Paper 11.

15. V. Foldyna, A. Jakobova and Z. Kubon: 'Assessment of Creep Resistance of 9–12% Cr Steels with respect to strengthening and degradation processes', *Materials Ageing and Component Life Extension*, Conf. Milano, 10–13 October 1995.

16. F. Hofer and P. Warbichler: *Prakt. Met. Sonderbd.*, 1995, **26**, 393.

17. Z. Kubon and V. Foldyna: 'The effect of Nb, V, N and Al on the creep rupture strength of 9–12%Cr-steels', *Proc. 17th VDEh Conference Langzeitverhalten warmfester Stähle und Hochtemperaturwerkstoffe*, Düsseldorf, 1994.

Thermodynamic Prediction of Microstructure

J. HALD[1] AND Z. KUBOŇ[2]

[1] *Department of Metallurgy, Technical University of Denmark 204, DK-2800 Lyngby, Denmark.*

[2] *VÍTKOVICE, j. s. c. Research Institute, Pohraniční 31, CZ-706 02 Ostrava, Czech Republic.*

ABSTRACT

In recent years the quality of thermodynamic equilibrium calculations on personal computers has been increased to a very high level by development of validated databases of thermochemical data. This is demonstrated by examples for 9–12%Cr power plant steels, including calculations of amount of δ-ferrite, transformation temperatures and equilibrium phases in new alloys like P91 and NF616. By combining the equilibrium calculations with kinetic models a description of Laves Phase precipitation in steel NF616 was achieved, and creep tests have indicated that Laves Phases produce significant precipitation strengthening in this steel. It is expected that thermodynamic equilibrium calculation will play an important role in the further development of the 9–12%Cr steels.

1. INTRODUCTION

Thermodynamic calculation of equilibrium phases in alloys as a function of solute content and temperature has been an important topic in metallurgy for decades. But for alloy systems containing more than a few elements the governing equations for thermodynamic models become far too complicated to solve manually within reasonable timescales, and this has limited the practical application of thermodynamics in metallurgy. In recent years this situation has changed. With the development of computer packages relying on validated databases, and with the increased capacity and calculation speed of personal computers (PCs) thermodynamic equilibrium calculation of complex alloy systems on PC is now readily available to many metallurgists. With such computer packages it is possible to calculate the amount and composition of equilibrium phases in alloys with up to 20 elements just by entering the chemical composition. In this paper results achieved for 9–12%Cr steels with such computer packages are demonstrated by examples.

1.1 The database

As mentioned the computer packages for equilibrium calculations rely on validated databases. These databases contain experimental thermochemical data for single elements and phases as a function of chemical composition, temperature and pressure. The development of such databases in recent years to contain more information and more accurate information is the strength of the computer packages, but at the same time the databases set the limitations to the achievable accuracy of the results. In Europe development of thermodynamic databases is undertaken by the SGTE (Scientific Group Thermodata Europe), which is an international consortium of seven European research centres.[1] The work of this group has resulted in the SGTE database containing assessed data for single elements and for condensed phases in a large number of binary, ternary and higher order alloy systems. The SGTE database is continuously upgraded with several new alloy systems each year.

Calculations of equilibria with the computer packages are made by combining the informations from the assessed systems in the database and extrapolating into the systems decribed by the user. The thermochemical data are stored as empirical mathematical functions of temperature and pressure, and equilibrium is calculated by minimising the Gibbs free energy of the system.

The user needs to define the amount of matter, the chemical composition (including up to 20 elements), the temperature and the pressure, before equilibrium can be calculated. Routines in the computer packages especially useful for condensed alloy systems, allow calculations of variations in equilibria as a function of temperature or solute content. Other possibilities with the packages include equilibria between solids and gases, and solidification simulations. Examples in this paper of equilibrium calculations for 9–12%Cr steels were calculated with the 'Thermocalc' program version j using the SGTE database version 1991. 'Thermocalc' is developed at the Royal Institute of Technology in Stockholm.[2]

1.2 The 9–12%Cr steels

The class of 9–12%Cr alloys discussed in this paper are tempered martensite steels alloyed with strong carbide formers like Mo, W, V and Nb for high creep strength at temperatures up to around 600°C. These alloys were developed for use as forged components for steam turbines and for steam pipework in fossil fired power plants. In recent years these 9–12%Cr steels have received great attention since alloy development have clearly improved their creep strength. Use of the newly developed 9–12%Cr steels have lead to significantly increased steam pressure and temperature and thereby to improved thermal efficiency of fossil fired power plants.[3]

The 9–12%Cr steels are normally delivered after a final heat treatment consisting of normalising and tempering. Normalising is done at a temperature, where the steels are austenitic (around 1050°C), and a basic characteristic of the 9–12%Cr steels is their ability to air harden to martensite in section thicknesses up to approximately 100 mm. For larger components oil quench-

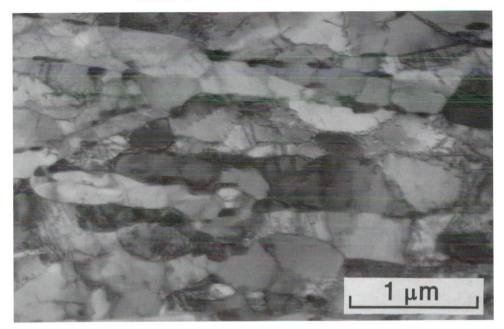

Fig. 1   Thin foil of steel X20CrMoV121.

ing must be used to obtain through hardening. Tempering is carried out in the range 650°C to 780°C. The lower tempering temperatures are used for components such as turbine rotors, where high tensile strength is required, whereas the higher tempering temperatures are used for pipework where high ductility and tempering resistance during post weld heat treatment are essential.

The normalising and tempering heat treatment leads to a microstructure of tempered martensite consisting of fine ferrite subgrains with carbides precipitated on subgrain boundaries and on prior austenite grain boundaries, Fig. 1. A dense distribution of fine carbides with high thermal stability in the 9–12%Cr steels produces high creep resistance, which is the main property required for this class of steels.

2. EXAMPLES

2.1 δ-ferrite and transformation temperatures.
In order for the 9–12%Cr steels to form 100% tempered martensite they need to be transformed to a fully austenitic microstructure during normalising. The ability to form a fully austenitic microstructure depends on the composition balance between the austenite forming alloying elements Ni, Mn, Cu, Co, C and N and the ferrite forming alloying elements Cr, Si, Mo, W, V and Nb. As the compositions of the 9–12%Cr steels tend to be close to the borderline for δ-ferrite stability imbalance between austenite and ferrite forming

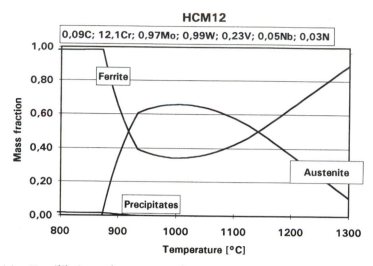

Fig. 2(a)   Equilibrium phases in steel HCM12.

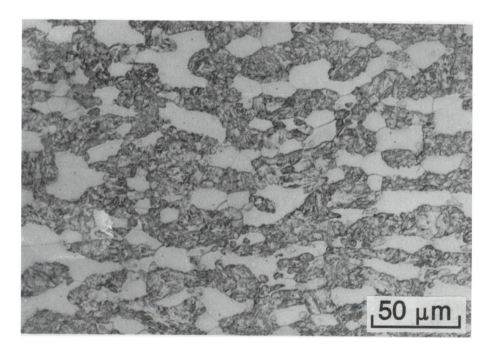

Fig. 2(b)   Microstructure in steel HCM12 after heat treatment 1050°C/1h/AC + 780°C/1 h/AC.

Table 1   Calculated and measured transformation temperatures for three 9–12%Cr steels

Steel	Calc. A_{e1}	Meas. A_{c1}
X20CrMoV121	787°C	815°C
HCM12A	766°C	805°C
NF616	821°C	840°C

elements will stabilise δ-ferrite at high temperature and reduce hardenability and forgeability of the steels. This is seen in the alloys EM12[4] and HCM12,[5] which because of high δ-ferrite content and reduced forgeability, can only be produced in form of small tubes.

An approximate assessment of the proper balancing of ferrite and austenite forming elements has traditionally been obtained from the well known Schaeffler or Schneider diagrams.[6] But with thermodynamic equilibrium calculations a much more accurate assessment of δ-ferrite stability at normalising temperatures is now possible. In Fig. 2a such a calculation is shown for the steel HCM12 containing approximately 30% δ-ferrite. In Fig. 2b a very good coincidence between the calculated amount of δ-ferrite at 1050°C and the observed amount in the steel can be observed.

Figure 2a also shows the A_{e1} temperature, which is the lowest temperature for austenite stability. A comparison between measured A_{c1} transformation temperatures and A_{e1} temperatures calculated with 'Thermocalc' for some 9–12%Cr steels are shown in Table 1. In general the calculated temperatures are lower than the measured values. This discrepancy is explained by the fact that calculated temperatures are equilibrium transformation temperatures, whereas the measured values are generally obtained by dilatometric measurements during heating, which introduce supersaturation effects. A German study have shown that with a heating rate of 5°C min.$^{-1}$ measured A_{c1} temperatures for 12CrMoV steels are approximately 30°C higher than A_{e1} temperatures determined by isothermal heating.[7] With a correction of 30°C on the measured values in Table 1, good coincidence between the measured and calculated transformation temperatures is observed.

2.2 Tempering of a 12%Cr steel
During final tempering of the 9–12%Cr steels the martensite is softened, and carbides and nitrides precipitate in the steels in well known precipitation sequences with increasing tempering temperature or tempering time. The precipitation sequences depend on steel composition. In a pure 0.1%C–12%Cr–0.02%N the sequence is:

$$M_3C \rightarrow \{M_2X + M_7C_3\} \rightarrow M_{23}C_6 + M_2X$$

Table 2 Chemical composition (wt.%), heat treatment and creep rupture strength of three pipe steels

	X20CrMoV121	P91	NF616
Heat treatment	1050°C/1h/AC +750°C/2h	1050°C/1h/AC +780°C/1h	1065°C/2h/AC +770°C/2h
C	0.20	0.10	0.106
Si	0.24	0.36	0.04
Mn	0.47	0.37	0.46
P	0.026	0.011	0.008
S	0.009	0.001	0.001
Cr	11.59	8.30	8.96
Mo	0.98	0.95	0.47
W	–	–	1.84
V	0.28	0.21	0.20
Nb	–	0.07	0.069
B	–	–	0.001
N	0.032	0.053	0.051
Ni	0.39	0.15	0.06
Al	–	–	0.007
σB, 100 000 h, 600°C	59 MPa	94 MPa	132 MPa

where M_2X is based on Cr_2N.[8] Alloying with vanadium tends to suppress M_7C_3 formation, and the equilibrium precipitates are then $M_{23}C_6$ and MX, based on VN,[8,9] i.e:

$$M_3C \rightarrow \{M_2X + M_{23}C_6\} \rightarrow M_{23}C_6 + MX$$

This precipitation sequence was observed by Petri *et al.*[9] during tempering experiments with the 12CrMoV steel X20CrMoV121. In Fig. 3a the calculated equilibrium phases for this steel is shown. The chemical composition is shown in Table 2. For the equilibrium calculations all elements except P and S were entered. It is seen that the calculations correctly predict $M_{23}C_6$ and MX as equilibrium precipitates at temperatures below A_{e1}.

With thermodynamic equilibrium calculations it is in principle only possible to predict equilibrium phases, but a routine in the 'Thermocalc' program allows for suspension of the precipitation of phases from the calculations. After suspension of the $M_{23}C_6$ and MX precipitates the calculations were repeated, and the results now show M_3C, M_6C, M_7C_3 and M_2X as equilibrium precipitates below the A_{e1} temperature, Fig. 3b. The calculated result that M_3C and M_2X precipitate at low temperature is a qualitatively good predic-

X20CrMoV121

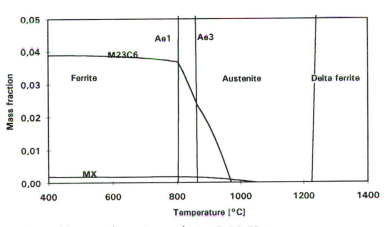

Fig. 3(a)   Equilibrium phases in steel X20CrMoV121.

X20CrMoV121- M23C6 and MX suspended

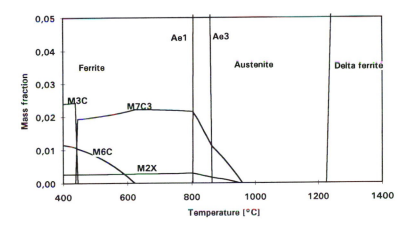

Fig. 3(b)   Equilibrium phase calculation in steel X20CrMoV121 after suspending phases M23C6 and MX.

tion of the observed precipitation sequence. The precipitation sequences occuring in 9–12%Cr steels are primarily determined by the diffusivities of interstitial and substitutional alloying elements at the tempering temperature, so in order to obtain a full quantitative description of preciptation sequences, kinetic models involving nucleation theory and diffusion effects should be applied.[10] But it is interesting to note that a qualitative indication of precipitation sequences can be obtained with the equilibrium calculations.

Besides the amount and type of precipitates, the thermodynamic calculations shown in Fig. 3a also produce the equilibrium chemical compositions of

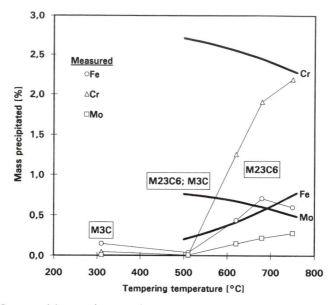

Fig. 4   Composition and type of precipitate in steel X20CrMoV121 after two hour tempering at different temperatures.[9] Full lines are the calculated composition of M23C6 carbides.

all phases as a function of temperature. This is very useful for the assessment of kinetics of the precipitation reactions. Figure 4 shows a comparison between calculated equilibrium amounts of Fe, Cr and Mo precipitated based on $M_{23}C_6$ and measured amounts of Fe, Cr and Mo precipitated after two hour tempering at different temperatures of the steel X20CrMoV121.[9] After tempering 750°C/2 h there is good agreement between the measurements and the calculations, which means that the $M_{23}C_6$ carbides reach equilibrium composition during tempering at this temperature. At lower temperatures the carbides can not reach equilibrium composition within two hours of tempering because diffusion of the substitutional alloying elements is too slow. Investigations after prolonged service exposure of X20CrMoV121 in steam pipes at 540°C have shown that the $M_{23}C_6$ carbides, which precipitated during final tempering at 750°C with relatively low Cr content, pick up Cr from the matrix during service exposure, so their chemical compositions approach the equilibrium composition at 540°C predicted by the thermodynamic calculations.

2.3 Precipitates in new 9–12% Cr steels
As mentioned the creep strength of the 9–12% Cr steels have been improved during the past 10–15 years by intensive alloy development. The alloy

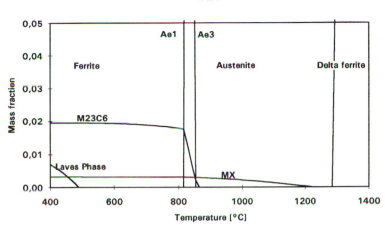

Fig. 5(a) Equilibrium phases in steel P91.

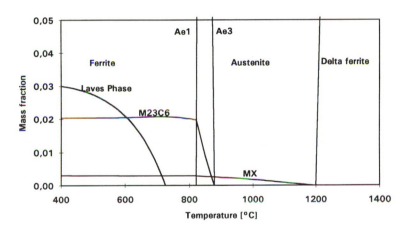

Fig. 5(b) Equilibrium phases in steel NF616.

changes which have been made are illustrated by the development of two new steam pipe steels, for which the chemical compositions, heat treatments and creep rupture strengths are shown together with a traditional 12CrMoV steel in Table 2. The 12CrMoV steel X20CrMoV121 has been used in continental Europe for steam pipes since the late 1950s.[11] The 9Cr–Mo steel P91 was developed in the late 1970s by Oak Ridge National Laboratories and Combustion Engineering in USA,[12] and it is now being introduced in power plants with advanced steam parameters. The 9Cr–W steel NF616 was invented by Prof. T. Fujita at Tokyo University in the middle of the 1980s, and it has in recent years been developed in cooperation with Nippon Steel Corporation to a stage where it is now ready for commercial application in power plants with further advanced steam parameters.[13]

As seen from Table 2 the alloy changes which have been introduced in P91 and NF616 are: balanced alloying with V, Nb and N and for NF616 alloying with W. Microstructural investigations have revealed that these alloy changes have mainly resulted in a change of the type and distribution of precipitates.[14] Figures 3a, 5a and 5b show the equilibrium phases in X20CrMoV121, P91 and NF616 calculated with 'Thermocalc'.

The main precipitate in the 9–12%Cr steels is the $M_{23}C_6$ carbide consisting of Cr, Fe Mo, (W) and C. This carbide produces the basic creep strength of X20CrMoV121 by precipitating on subgrain boundaries during tempering, Fig. 1. The $M_{23}C_6$ carbides increase creep strength by retarding subgrain growth, which is a major source of creep strain in these alloys.[15,16]

When comparing the chemical compositions of X20CrMoV121 and P91, Table 2, and the amount of $M_{23}C_6$ produced in the steels, Figs 3a and 5a, it may seem strange that the creep rupture strength is improved by approximately 50% at 600°C by decreasing the amount of strengthening $M_{23}C_6$ carbides. This is explained by the MX precipitates which are introduced in P91 by well balanced alloying with V, Nb and N.

The MX precipitates consist mainly of V, Nb and N, and they are found to precipitate within subgrains, where they pin down free dislocations and in that way increase creep strength, Fig. 6. The thermal stability of the MX precipitates is very high, and that can explain their tremendous effect on creep strength. It is interesting to note that the equilibrium calculations indicate that the strengthening effect of Vanadium in X20CrMoV121 is obtained by precipitation of Vanadium nitrides. This means that even though Nitrogen is not a specified alloying element in X20CrMoV121, the tramp Nitrogen content contributes to high creep strength of this steel.

From Figs 3a and 5 it is clear that MX is a stable phase at 1050°C in P91 and NF616, whereas in X20CrMoV121 it is dissolved at 1050°C. This means that some MX will remain undissolved during normalisation of P91 and NF616. It has been shown that these undissolved particles are effective in preventing grain growth during normalising, and in that way they contribute to improved toughness of the new steels. 'Thermocalc' calculations show that the undissolved MX at 1050°C are carbonitrides containing V, Nb, Cr, N and C, see Table 3. The Nitrogen remaining in solid solution during austenitisation is available during tempering for the fine MX precipitation within subgrains, which contribute to the increased creep strength. The calculations show that the fine MX precipitating during tempering are carbonitrides with high Vanadium content, Table 3.

The alloying change from P91 to NF616 is primarily a reduction of the Mo

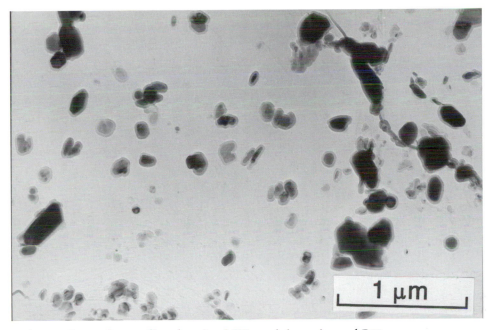

Fig. 6 Extraction replica showing MX precipitates in steel P91.

Table 3 MX compositions (atomic %) in steel NF616 calculated by 'Thermocalc'

NF616	1050°C	770°C
C	6.67	1.41
Cr	1.28	0.25
N	41.67	46.86
Nb	17.13	3.12
V	33.07	47.74

content and addition of W. The effect of these changes is an approximately 30% increase in creep rupture strength. The 'Thermocalc' calculations indicate that tne changes in Mo and W contents will lead to the formation of intermetallic Laves Phase $(Fe,Cr)_2(Mo,W)$ at temperatures below app. 700°C in NF616, Fig. 5b. This means that no Laves Phase is found in the steel in the as delivered condition, but Laves Phase will precipitate during service exposure in power plant at temperatures around 600°C. Microstructural investigations of NF616 isothermally aged at 600°C and 650°C have confirmed the Laves Phase precipitation,[17] Fig. 7. In P91 Laves Phase has been observed after isothermal ageing at 600°C but not after ageing at 650°C.[17] This is not in agreement with the 'Thermocalc' calculations which predict Laves Phase only

Fig. 7   Thin foil showing Laves Phase in steel NF616 after ageing at 600°C.

to be stable below 489°C, Fig. 5a. Inspection by the authors of the SGTA database has led to the conclusion that the model describing Laves Phase in Mo-alloyed steels is not sufficiently assessed for 9%Cr–Mo steels.

The examples for 9–12%Cr steels shown above demonstrate the high accuracy of thermodynamic equilibrium calculations, which has been achieved by intensive development of the SGTA database in recent years. But at the same time the inaccuracies of the Laves Phase predictions in 9Cr–Mo steels show that further upgrading of the database is still necessary. This further demonstrates that even though a very high level of accuracy has been reached, blind uncritical adoption of all results from the computations without any experimental verification should be avoided.

2.4 Kinetic modelling

It has long been claimed that the creep strengthening effect of Mo and W in 9–12%Cr steels is by solid solution hardening, and therefore Laves Phase precipitation removing Mo and W from solid solution should be regarded as detrimental to creep strength of the new steels. It is therefore of great importance for assessment of the long term stability of the new W alloyed 9Cr steels to know the kinetics of the Laves Phase precipitation.

The equilibrium calculations give no information about the kinetics of such precipitation reactions, but as shown above equilibrium is very nearly reached during final tempering of the 9–12%Cr steels at 750°C or higher temperatures. This means that the equilibrium calculations produce the start point and the end point for the Laves Phase precipitation reaction taking place at temperatures below approximately 700°C. By combination of the equilibrium calculations with measurements of the amount of precipitates as a function of time and temperature, and with the Johnson-Mehl-Avrami equations, which empirically describe precipitation reactions, a full description of the overall precipitation of Laves Phase in the steel NF616 has been achieved:[17]

$$w(t) = 1 - \exp(-(t/\tau)^n) \tag{1}$$

For $n = 3/2$:

$$1/\tau = 2D \left(\frac{4}{3} \pi N \right)^{\frac{2}{3}} \left\{ \frac{C_0 - C_e^{\alpha}}{C_e^{\beta} - C_e^{\alpha}} \right\}^{\frac{1}{3}} \tag{2}$$

and:

$$D = D_o \exp\left\{ \frac{-Q_o}{RT} \right\} \tag{3}$$

Where:

$w(t)$ = Relative amount of Tungsten precipitated at time t.

C_0 = Equilibrium Tungsten concentration in Ferrite after tempering at 770°C, (1,615 wt%), calculated by Thermocalc.

C_e^{α} = $f_1(T)$, equilibrium Tungsten concentration in Ferrite at temperature T, calculated by Thermocalc.

C_e^{β} = $f_2(T)$, equilibrium Tungsten concentration in Laves Phase at temperature T, calculated by Thermocalc.

N = Nucleation site number.

Q_0 = $55.1.10^3$ cal mole^{-1}.[18]

D_0 = 0.29 cm^{2}s^{-1}.[18]

R = 1.9872 cal mole^{-1} k^{-1}

Empirical fits to eqn (1) giving n and $1/\tau$ at 600°C and the assumptions that N is independent of temperature in the interesting range, and that eqn (2) holds also for the experimentally determined n, allowed calculations of time-temperature curves and the time-temperature-precipitation diagram for Laves Phase precipitation in NF616, Figs. 8 and 9. It is seen that the calculations give a good description of the experimental results.

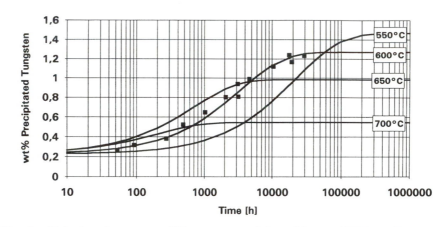

Fig. 8 Calculated amount of Tungsten precipitated in steel NF616. Squares are experimental results used to calibrate the model.

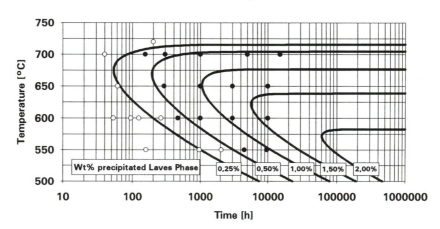

Fig. 9 Calculated Time-Temperature-Precipitation diagram for Laves Phase in steel NF616. Open circles: No Laves Phase, and full circles: Laves Phase detected by experiment.

Creep tests were made on NF616 in different heat treatment conditions, Fig. 10. The material was tested in three conditions: (1) the as received condition, (2) after ageing 650°C/10 000 h and (3) after ageing 720°C/200 h. Heat treatment conditions (2) and (3) are similar when assessed by the ageing parameter $P = T(20 + \log t)$, where T is temperature in K and t is time in hours, and materials (2) and (3) had similar tensile strength. Laves Phase precipi-

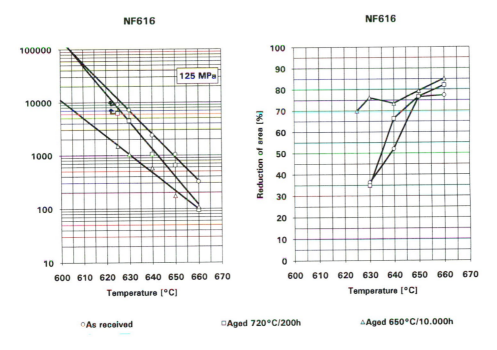

NF616   **NF616**

Temperature [°C]   **Temperature [°C]**

○As received   □Aged 720°C/200h   △Aged 650°C/10.000h

Fig. 10   Creep rupture tests of steel NF616 after ageing.

tation in material (2) has finished before creep testing, Fig. 9, and no Laves Phase has precipitated in material (3) prior to creep testing, Fig. 10. The fact that no difference in rupture life is observed between materials (2) and (3) in the short term test, during which no Laves Phase precipitates, show that there is only small effect of Mo and W in solid solution on creep strength. Material (2) has 0.6 wt% W in solid solution before creep testing, and material (3) has 1.6 wt% W in solid solution before creep testing.

During the longer lasting creep tests at lower temperature Laves Phase precipitate in material (3) but not in material (2), where Laves Phase precipitation was exhausted before creep testing. The fact that material (3) is now stronger than material (2) is a strong indication of precipitation strengthening by Laves Phase. This is supported by the area reduction measurements, which show that in materials (1) and (3) where Laves Phase precipitates during creep the ductility is lowered, whereas it stays high in material (2), Fig. 10.

These results indicate that Laves Phase precipitation during creep should be regarded as the effect producing increased creep strength of steel NF616 compared with P91. Laves Phase precipitation should therefore be regarded as a beneficial effect in the 9–12%Cr steels if it does not impair ductility. It is known that Laves Phases coarsen relatively rapidly and the long term creep stability of steels like NF616 relying part of their creep strength on Laves

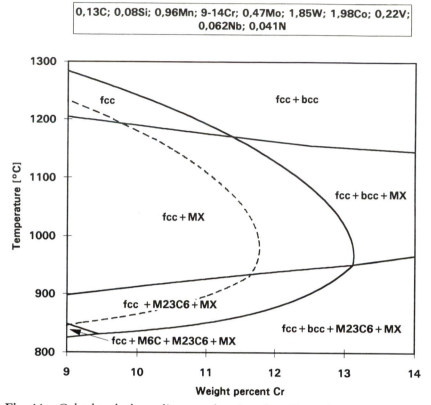

Fig. 11 Calculated phase diagram showing the effect of Cr in an experimental steel. The dotted line show the ferrite phase boundary calculated without Co addition.

Phase precipitation should therefore be regarded with some caution. Once all Laves Phase has precipitated and coarsened, the creep strength is bound to come down, but if this does not happen until after 300 000 hours at service temperature the steels can be used with confidence in power plants. All test results seen so far including creep tests at 600°C on NF616 with running times approaching 60 000 hours have been encouraging in this respect.

2.5 Alloy development

It has been demonstrated above that the accuracy of thermodynamic calculations have now reached such a high level that it can be used to significantly increase our knowledge about the effect of alloying elements on the metallurgy of existing 9–12%Cr steels. The last example of such calculations to be presented in this paper points into the future. There is an ongoing quest for further development of the 9–12%Cr steels to even higher creep strength than the new alloys presented above, and in this context the thermodynamic equilibrium calculations can play an important role in the identification of new alloys to be tested.

A way to improve the creep strength of the 9–12%Cr steels could be to increase the volume fraction of precipitates in the steels by increasing the amount of alloying elements like Cr, Mo and W which form carbides, nitrides or intermetallic phases in the steels. This is however not simple since these alloying elements are ferrite formers which will stabilise δ-ferrite at the normalising temperature and impair forgeability. Addition of ferrite forming elements must therefore be balanced by the addition of austenite forming elements. The addition of the most effective austenite forming elements C and N must however be limited because they impair weldability. Ni addition must also be limited because it lowers the austenite transformation temperature, which makes tempering of the steels difficult. Research work has also indicated that Ni reduces the thermal stability of precipitates.[19] Cu addition as austenite stabiliser has recently been tried with success in the steel HCM12A, but Cu forms precipitates in ferrite and little is known about the stability of the Cu precipitates.[20] Another way forward which is currently being investigated is Co-addition. This element stabilises austenite without suppressing the austenite transformation temperature. In Fig. 11 the effect of Co addition to a Cr steel is shown with a phase diagram calculated by 'Thermocalc'. For this alloy it is interesting to increase the Cr content to improve steam oxidation resistance. It may be seen that the addition of approximately 2% Co allows for a simultaneous increase of the Cr content to just above 12% without stabilising δ-ferrite.

This demonstrates how thermodynamic equilibrium calculations can be used to study the effect of alloy changes on the amount of equilibrium phases, and thereby assist in the selection and balancing of the chemical compositions of new alloys to be produced and tested. Much more complicated models are needed to fulfil the metallurgists dream to be able to predict the influence of alloy changes on mechanical properties like e.g. creep strength, but the introduction of thermodynamic equilibrium calculations is a very important step in this direction.

3. SUMMARY

The quality of thermodynamic equilibrium phase calculations on personal computers has been increased to a very high level in recent years by the development of validated databases of thermochemical data. By examples it has been demonstrated how such calculations can significantly increase our understanding of the metallurgy of the 9–12%Cr ferritic power plant steels.

Examples include the calculation of δ-ferrite stability and austenite transformation temperatures, and of equilibrium phases and precipitate composition in a 12CrMoV steel.

It has been shown how equilibrium calculations can aid in the understanding of strengthening phases in newly developed 9%Cr alloys like the P91 and NF616 with high creep strength up to temperatures around 600°C.

By combination of the equilibrium calculations with empirical kinetic models and experimental data the precipitation of intermetallic Laves Phases in steel NF616 was modelled. Creep tests on the steel in different heat treatment conditions have indicated that the discredited Laves Phases should really be regarded as creep strengthening agents in the 9–12%Cr steels.

It is expected that thermodynamic equilibrium calculations will be an important tool for the further development of new 9–12%Cr steels with even higher creep strength.

ACKNOWLEDGMENTS

The authors wish to acknowledge the financial support from the COST management committee, which allowed Z. Kuboň to stay at the Department of Metallurgy in Lyngby on a scientific mission during March 1995.

REFERENCES.

1. A.T. Dinsdale: *Calphad*, 1991, **15** (4), 317–425.
2. B. Sundman, B. Jansson and J.-O. Andersson: *Calphad*, 1985, **9** (2), 153–190.
3. R. Blum, J. Hald, W. Bendick, A. Rosselet and J.C. Vaillant: *VGB Kraftwerkstechnik*, 1994, **74** (9), 641–652.
4. M. Caubo and J. Mathonet: *Revue de metallurgie*, May 1969, 345–360.
5. A. Iseda, Y. Sawaragi, H. Teranishi, M. Kubota and Y. Hayase: *The Sumitomo Search*, 1989, **40**, 41–56.
6. M. Schirra: *Stahl und Eisen*, 1992, **112** (10), 117–120.
7. G. Kalwa and E. Schnabel: *Proc. Conf. Werkstoffe und Schweisstechnik im Kraftwerk 1989*, VGB, Essen, 1989, 162–185.
8. K.J. Irvine, D.J. Crowe and F.B. Pickering: *J. Iron and Steel Inst.* 1960, **195**, 386–405.
9. R. Petri, E. Schnabel and P. Schwaab: *Arch. Eisenhüttenwes.* 1981, **52** (1), 27–32.
10. R.C. Thomson and H.K.D.H. Bhadeshia; *Metall. Trans. A*, 1992, **23A**, 1171–1179.
11. H. Jesper and H.R. Kautz: *Proc. Conf. Werkstoffe und Schweisstechnik im Kraftwerk 1985*, VGB, Essen, 1985, 274–316.
12. V. K. Sikka: *Proc. Conf. Topical Conference on Ferritic Alloys for use in Nuclear Energy Technologies,* J.W. Davies and D.J. Michel eds, The Metallurgical Society of AIME, Pennsylvania, 1984, 317–327.
13. H. Naoi, H. Mimura, M. Oghami, H. Morimoto, T. Tanaka, Y. Yazaki

and T. Fujita: *Proc. Conf. 'New Steels for Advanced Plant up to 620°C'*, E. Metcalfe ed. London, May 1995, EPRI, USA 1995, 8–29.

14. J. Hald: *TEM Investigations in New 9–12%Cr Steels for High Temperature Applications*, report from ELSAM/ELKRAFT/Department of Metallurgy, Technical University of Denmark, April 1988.

15. G. Eggeler, N. Nilsvang and B. Ilschner: *Steel Research*, 1987, **58** (2), 97–103.

16. S. Straub: *Verformungsverhalten und Mikrostruktur warmfester martensitischer 12% Chromstähle*, VDI Verlag, Düsseldorf, 1995.

17. J. Hald: *Proc. Conf. New Steels for Advanced Plant up to 620°C*, E. Metcalfe ed., London, May 1995, EPRI, USA, 1995, 152–173.

18. *CRC Handbook of Chemistry and Physics*, R.C. Weast and M.J. Astle eds, F–54, CRC Press, Florida, 1982–1983.

19. A. Strang and V. Vodarek: *Proc. Conf. Microstructural Development and Stability in High Chromium Ferritic Power Plant Steels*, June 1995, Institute of Materials and Cambridge University, 1995.

20. Y. Sawaragi, A. Iseda, K. Ogawa, F. Masuyama and T. Yokoyama: *Proc. Conf. 'New Steels for Advanced Plant up to 620°C'* E. Metcalfe ed., London, May 1995, EPRI, USA, 1995, 45–55.

Modelling the Development of Microstructure in Power Plant Steels

J.D. ROBSON AND H.K.D.H. BHADESHIA

University of Cambridge, Department of Materials Science and Metallurgy, Pembroke Street, Cambridge CB2 3QZ, UK

ABSTRACT

The quantitative design of alloys is now a substantial subject in metallurgy. There are numerous software packages available, which when presented with appropriate thermodynamic data, enable routinely the calculation of phase diagrams as a function of variables such as temperature, pressure and solute content. This thermodynamic base can sometimes be used to formulate kinetic models which allow the prediction of useful microstructures and mechanical properties. This paper describes such work in the context of critical alloys used in the power generation industries.

INTRODUCTION

Steels used in the manufacture of power plant range from those designed to resist creep deformation at temperatures around 600°C, to others which are exposed to relatively low temperatures where the primary design criterion is toughness.[1,2] The microstructures of power plant alloys often consist of δ-ferrite, martensite, bainite, allotriomorphic ferrite and retained austenite as the major phases obtained following a normalising heat-treatment. However, these microstructures are then subjected to very severe tempering ($\simeq$700°C, several hours) causing general coarsening and the precipitation of alloy carbides. There may be a total of eleven separate heat-treatments during the manufacture of a rotor. Subsequent service at elevated temperatures ($\simeq$600°C) causes further microstructural changes over the 30 year lifetime of the plant. Some of the metallurgical issues are summarised in Fig. 1.

Given this complexity, it is not surprising that the vast majority of commercial power plant alloys have been designed using accumulated experience and great skill. Any attempt to model this design process must recognise the full complexity of the microstructure and the properties that depend on it.[3] Basic science is not yet ready to tackle all the necessary problems. A range of methods must therefore be used; these are introduced below; the purpose of this paper is to illustrate progress in the quantitative design of power plant steels.

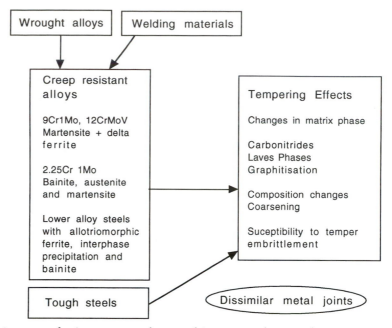

Fig. 1 Aspects of microstructural control in power plant steels.

MODELLING METHODS

Blind procedures such as regression[4] or neural network analysis[5] can reveal new regularities in data. They closely mimic human experience and are capable of learning or being trained to recognise the correct science rather than nonsensical trends. Unlike human experience, these models can be transferred readily between generations and developed continuously to make design tools of lasting value. Modelling also imposes a discipline on the digital storage of valuable experimental data, which may otherwise be lost with the passage of time.[6]

The ideal models are those based on firm physical principles. Once established, they can be used with greater confidence and are capable of predicting entirely new phenomena. In materials science any attempt to model by simplification (i.e. convert into a pure problem) is likely to diminish the value of the model to technology. A practical and useful method is always one which is a compromise between basic science and empiricism.[7] Good engineering has the responsibility to reach objectives in a cost and time-effective way.[8]

The modelling approach to the design of materials and processes is important and in great demand by industry because empirical experiments are now too expensive. In a competitive environment, there may also be severe time penalties. Good modelling techniques can reduce the time from conception to production, can provide quantitative tools of lasting value, and permit a reliable and easy route for the transfer of technology between university and industry.

A design process must address all aspects in a connected way.[3] This means that each aspect of manufacture must be made amenable to modelling. The presentation that follows is, in this respect, unsatisfactory because all the necessary mechanical properties are not addressed. An attempt has nevertheless been made to demonstrate the principles involved, both for properties which lend themselves to 'rigorous' treatment and others which do not.

For the purpose of this paper we shall concentrate on the development of physical models only, beginning with calculations based on thermodynamics.

THERMODYNAMICS AND DEVIATIONS FROM EQUILIBRIUM

Pippard[9] defined a system to be in equilibrium when there is no perceptible change no matter how long one waits. This definition emphasises the limitation of equilibrium phase diagram calculations in the design of alloys. After all, most useful microstructures are far from equilibrium. For example, the strength and toughness can be optimised simultaneously by refining the grain structure, i.e. by deviating from equilibrium. Even though power plant steels have quite stable microstructures when installed, they change substantially during long term service. Stability only has meaning when considered in the context of service. The tempered steels are not at equilibrium over a period of 30 years at 600°C.

Pippard's concept of equilibrium is not practical. Metastable equilibrium is more appropriate. It is therefore convenient that all thermodynamic principles continue to apply to the metastable state. The pioneering work by Kaufman, Hillert, Kubachewski, and others has made it possible to calculate, using thermodynamic functions and data, the state of metastable equilibrium between many phases in a large number of systems, each of which might contain several components.[10-14] This is particularly the case for iron alloys, which is good because they are so useful.

These phase diagram calculations are used routinely in industry and academia. For example, Laves phases are generally believed to be detrimental to the creep strength of alloy steels. They may be intrinsically brittle, but their effect on creep is probably better explained by the fact that their precipitation causes a depletion of solute from the ferritic matrix. The consequent loss of strength is not compensated by the precipitation of Laves phases because they are incoherent and coarse. The depletion caused by Laves phase formation is illustrated using a calculated phase diagram in Fig. 2.

Of course, it remains to be demonstrated quantitatively that the loss of strength due to solute depletion is not compensated by precipitation hardening due to Laves phase. Some data for the solid solution strengthening contributions from different elements are presented in Table 1:[6,7,17] note that

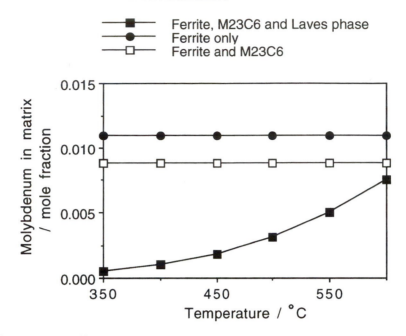

Fig. 2 An illustration of the change in the dissolved molybdenum concentration when Laves phase precipitates in ferrite. The calculations have been carried out using the MTDATA program[15] and SGTE database for iron.

whereas the strength of pure iron increases as the temperature is reduced, strengthening due to substitutional solutes often goes through a maximum as a function of temperature. Indeed, there is some solution *softening* at low temperatures because the presence of a foreign atom locally assists a dislocation to overcome the exceptionally large Peierls barrier of body-centered cubic iron, at low temperatures.

It is emphasised here that the data presented in Table 1 are 'pure', in the sense that they are derived from very careful experiments in which the individual contributions can be measured independently. There exist a vast range of published equations in which the strength is expressed as a (usually linear) function of the weld chemistry [see for example, Table 5.2 of reference 18]. These equations are derived by fitting empirically to experimental data and consequently include much more than just solid solution strengthening components.

CONSTRAINED EQUILIBRIA

Calculations of the equilibrium state are obviously of limited value because equilibrium takes forever to achieve in practice. Nevertheless, the same methodology can be used to examine *constrained* equilibria. Kinetic factors often prevent transformations from occurring under equilibrium conditions; Gibbs,[19] Darken,[20] Darken and Gurry,[21] Baker and Cahn[22] and Cahn[23] have

Table 1   Strength (MPa) of pure iron as a function of temperature, and solid solution strengthening terms for ferrite, for 1 wt.% of solute. The data are for a strain rate of $0.0025 \, \text{s}^{-1}$.

	200°C	100°C	Room Temperature (23°C)	−40°C	−60°C
Fe	215	215	219	355	534
Si	78	95	105	70	−44
Mn	37	41	45	8	−57
Ni	19	23	37	−2	−41
Mo	–	–	18	–	–
Cr	7.8	5.9	5.8	7.4	15.5
V	–	–	4.5	–	–
Co	1.0	1.8	4.9	9.1	5.8

discussed the different kinds of kinetically constrained equilibria that arise naturally and which can be treated using thermodynamics alone.

One such example is when a transformation is so rapid that one or more of the components cannot redistribute among the phases in the time scale of the experiment. In steels the diffusion coefficients of substitutional and interstitial components differ so greatly that there are circumstances where the sluggish substitutional alloying elements may not have time to redistribute during transformation of austenite to ferrite, even though carbon may partition into the austenite.[24–26]

Hultgren[24] introduced the term 'paraequilibrium' to describe this constrained equilibrium between two phases which are forced to have the same substitutional-solute to iron atom ratio, but which (subject to this constraint) achieve equilibrium with respect to carbon.

FERRITE

We now consider the thermodynamic definition of paraequilibrium for a Fe-X-C alloy, with Fe, X and C identified by the subscripts 3, 2 and 1 respectively. Ferrite is a component of many low-alloy power plant steels, so paraequilibrium is first discussed in terms of the austenite to ferrite transformation where a great deal of research has already been done. Equilibrium between austenite and ferrite (of homogeneous compositions $x_{i\gamma}$ and $x_{i\alpha}$ respectively, with $i = 1$, 2 or 3) is said to exist when

$$\mu_i^{\alpha} \{x_{1\alpha}, x_{2\alpha}, x_{3\alpha}\} = \mu_i^{\gamma} \{x_{1\gamma}, x_{2\gamma}, x_{3\gamma}\} \tag{1}$$

where μ represents the chemical potential of an element in a particular phase.† This is illustrated, for a fixed temperature, in Fig. 3 where the equilibrium condition is defined by a common tangent-plane. The intercepts on the pure component axes represent the values of the respective chemical po-

---

† The free energy of a solution depends on the contributions from each of its component elements, weighted according to their respective concentrations. The free energy of a particular element within that solution is called its chemical potential.

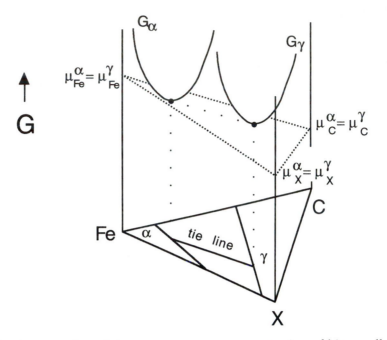

Fig. 3 An extension of the common-tangent construction of binary alloys, to a tangent plane construction for ternary alloys, for a fixed temperature. Each pair of points defined by the contact of a plane tangent to both the austenite and ferrite free energy surfaces defines a tie-line of an isothermal section of the Fe–C–X phase diagram. The tangent plane can be rocked, whilst maintaining contact with the two free energy surfaces, to generate the set of tie lines which defines the α + γ equilibrium phase field. The intercepts of the tangent plane with the vertical axes give the chemical potential of each component.

tentials of those components. There are no gradients of chemical potential when the system is at equilibrium, either within a given phase or across phase boundaries.

Austenite and ferrite are said to be in paraequilibrium when

$$x_{2\alpha}/x_{3\alpha} = x_{2\gamma}/x_{3\gamma} = \bar{x}_2/\bar{x}_3 \tag{2}$$

$$\mu_1^\alpha = \mu_1^\gamma \tag{3}$$

where x represents the average concentration in the alloy as a whole.

The Gibbs free energy change ΔG per mole reacted for reactions in a closed system, when an infinitesimal amount of material of composition $x_{i\alpha}$ is transferred from γ of composition $x_{i\gamma}$ to α of composition $x_{i\alpha}$ is given by:

$$\Delta G = x_{1\alpha}(\mu_1^\alpha - \mu_1^\gamma) + x_{2\alpha}(\mu_2^\alpha - \mu_2^\gamma) + x_{3\alpha}(\mu_3^\alpha - \mu_3^\gamma) \tag{4}$$

where the chemical potentials in the γ and α are evaluated at the compositions $x_{i\gamma}$ and $x_{i\alpha}$ respectively.

ΔG clearly equals zero when α and γ are in equilibrium, since the chemical potential of any component is uniform everywhere. The paraequilibrium state is also specified by setting $\Delta G = 0$, subject to the constraint of eqn 2, so that on combining eqns 2–4, we get:

$$\mu_2^\gamma - \mu_2^\alpha + (\mu_3^\gamma - \mu_3^\alpha)(\bar{x}_3 / \bar{x}_2) = 0 \qquad (5)$$

This equation is another description of the paraequilibrium state and was first derived by Gilmour et al.[27] The state of paraequilibrium is illustrated in Fig. 4 using tangent planes and the free energy surfaces of austenite and ferrite. There is now no *common* tangent plane, but instead two planes, each of which is only tangential to either the austenite or the ferrite free-energy surface. However, these two tangent planes have a common origin on the carbon axis, because the chemical potential of carbon must be identical in both phases during paraequilibrium. The chemical potentials of the other two elements are clearly not equal in austenite and ferrite since the tangent planes do not have common intersections on either the iron or X axes. For the case illustrated, the chemical potential of X is raised on transformation whereas that of Fe is reduced, in such a way that the two effects cancel to give a zero net free energy difference between austenite and ferrite. Equation 5 simply expresses this in a mathematical form.

Another important feature illustrated in Fig. 4 is that the two tangent planes share a line of intersection, which when projected onto the isothermal section of the phase diagram, gives the locus of all points along which there is a constant ratio of Fe/X, as required by the definition of paraequilibrium. Consistent with eqn 5, we see from the geometry illustrated in Fig. 4 that

$$(\mu_X^\gamma - \mu_X^\alpha)\bar{x}_X = (\mu_{Fe}^\alpha - \mu_{Fe}^\gamma)\bar{x}_{Fe}$$

Equation 5 can be generalised to define the paraequilibrium condition for a multicomponent system in which there are n mobile components and m immobile components.[28] by analogy with the ternary case, we have a uniform chemical potential for all the mobile components:

$$\mu_i^\alpha = \mu_i^\gamma \qquad \text{for } i = \text{ to } n \qquad (6)$$

and for the immobile components,

$$\sum_{j=1}^{m} \frac{\bar{x}_j}{\bar{x}_1} (\mu_j^\gamma - \mu_j^\alpha) = 0 \qquad \text{for } j = 1 \text{ to } m \qquad (7)$$

We see that during the equilibrium formation of α from γ, the chemical potential of carbon is identical in both phases at the interface, and this remains

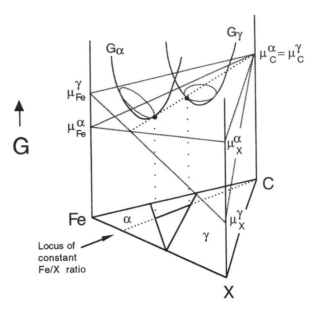

Fig. 4 The paraequilibrium state illustrated using the austenite and ferrite free energy surfaces for a substitutionally alloyed steel at a fixed temperature. There are two tangent planes, each plane being tangential to just one of the phases whilst cutting the free energy surface of the other phase. It is only the chemical potential of carbon which is identical in austenite and ferrite during paraequilibrium. Note also that the paraequilibrium tie line is parallel to a construction line radiating from the carbon corner of the isothermal section of the phase diagram. In addition, the α + γ two-phase field constricts to a point as the carbon concentration tends to zero.

the case when α forms from γ by a paraequilibrium mechanism. However, the equilibrium and paraequilibrium *concentrations* of carbon in α or in γ will in general differ because the chemical potential of carbon is a function of all elements in solution. The substitutional element concentrations are different for the two cases. The equilibrium phase diagram can not be used to specify interface tie-line compositions for paraequilibrium. Hillert[25] has shown that the paraequilibrium phase boundaries lie within the α + γ phase field of the equilibrium phase diagram. Furthermore, the tie-lines of the paraequilibrium α + γ phase field follow a path along which the substitutional to iron atom ratio is constant. Typical paraequilibrium and equilibrium Fe–Mn–C diagrams are illustrated in Fig. 5. It is clear that for any given alloy, a critical undercooling below the equilibrium transformation temperature is necessary before paraequilibrium transformation becomes feasible. This reflects the fact that a relatively lower free energy change accompanies the formation of ferrite which is forced to accept a non-equilibrium substitutional alloy content.

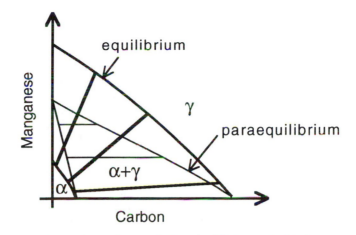

Fig. 5 Schematic isothermal sections of the equilibrium (bold lines) and paraequilibrium phase diagrams of the Fe–Mn–C system. The tie-lines for the paraequilibrium diagram are virtually horizontal since the Fe/Mn ratio is constant everywhere during paraequilibrium transformation.

PARAEQUILIBRIUM IN IRON-BASE CARBIDES

There is good evidence that some of the carbides which form during the tempering of martensite, or as a consequence of the bainite reaction, grow by a displacive mechanism.[29] Such a mechanism must naturally involve the diffusion of carbon, but not of substitutional solute or iron atoms, i.e., by a paraequilibrium mechanism. This is consistent with many data which show that the carbides associated with bainite and martensite do not partition substitutional solutes during transformation.[30–32] It is particularly interesting that the precipitation of cementite from martensite or lower bainite can occur under conditions where the diffusion rates of iron and substitutional atoms are incredibly small compared with the rate of precipitation (Fig. 6). The long-range diffusion of carbon atoms is of course necessary, but because of its interstitial character, substantial diffusion of carbon remains possible even at temperatures as low as $-60°C$. Thus, the formation of cementite in these circumstances must differ from the normal reconstructive decomposition reactions, which become sluggish at low temperatures. It has been believed for some time that the cementite lattice may be generated by the deformation of the ferrite lattice, combined with the necessary diffusion of carbon into the appropriate sites. The Fe/X ratio thus remains constant everywhere and subject to that constraint, the carbon achieves equality of chemical potential; the cementite is then said to grow by *paraequilibrium* transformation. The way in which the ferrite lattice could be deformed to produce the right arrangement of iron atoms needed to generate the cementite has been considered by Andrews[33] and Hume-Rothery *et al.*,[34] and the subject has been reviewed by

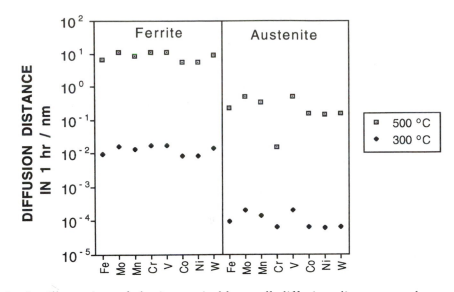

Fig. 6 An illustration of the inconceivably small diffusion distances at the temperatures where some iron carbides precipitate rapidly.

Yakel.[35] Further high-resolution evidence supporting the idea that the carbide particles grow by displacive transformation (involving the diffusion of just carbon) has been published most recently by Sandvik,[36] Nakamura and Nagakura[37] and Taylor *et al*.[38,39]

COMPOSITION CHANGES IN IRON-BASE CARBIDES

It can be demonstrated that for the early stages of carbide enrichment, the composition change during the isothermal annealing of a plate-shaped carbide particle which is surrounded by an infinite matrix is given by:[40]

$$t = \frac{\pi \left[x^\theta (\bar{c} - c^\theta) \right]^2}{16 D_\alpha (c^{\alpha\theta} - \bar{c})^2} \tag{8}$$

where t is the time at the annealing temperature, c^θ is the average substitutional solute content of the cementite (θ) at time t and x^θ is the thickness of the carbide plate. D_α is the diffusion coefficient of the substitutional solute in ferrite, $c^{\alpha\theta}$ is the concentration in ferrite which is in equilibrium with cementite, and c is the starting composition of the carbide (Fig. 7).

Therefore, the composition ($c^\theta - c$) should vary with time $t^{1/2}$ during isothermal annealing; this is a parabolic dependence, which contrasts with the frequent assumption in industry of $t^{1/3}$ kinetics. In practice, it is not really possible to distinguish between these two time exponents; the experimental data give reasonable linear fits with either relationship. Nevertheless, values

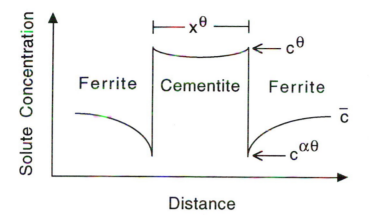

Fig. 7 The concentration profile that develops during the enrichment of a cementite particle during annealing.

of the diffusion coefficient extracted from such plots will only be correct for parabolic kinetics,[41] which can extrapolated with greater confidence than the $t^{1/3}$ relationship. Furthermore, the effect of the steel chemistry is explicitly included in eqn 8 via the equilibrium and average compositions, so that data from many different alloys can be interpreted together.

Equation 8 reveals a dependence of solute-enrichment on the particle size. Smaller particles should enrich faster and saturate earlier because they are smaller reservoirs for solute (Fig. 8a).

The assumption of an infinite matrix surrounding each particle, as used in deriving eqn 1, can easily be overcome using a finite difference model.[5] This method allows for soft-impingement, which is the overlap of the diffusion fields of adjacent particles. In addition, it predicts that the parabolic law will only be valid in the absence of significant soft-impingement. The enrichment kinetics are expected to slow down relative to the parabolic law after the onset of soft-impingement, until eventually the particle saturates at its equilibrium composition.

It has been emphasised that cementite often starts off with the same substitutional solute content as the alloy as a whole, but that the initial composition of pearlitic cementite cannot be predicted. It turns out that for calculation purposes, this is not a serious problem for long ageing times, since the degree of enrichment then tends to be much larger than the starting solute level (Fig. 8b).

COMPOSITION CHANGES IN Fe CARBIDES: ROLE OF CARBON

There are two effects which depend on the carbon concentration of the steel.

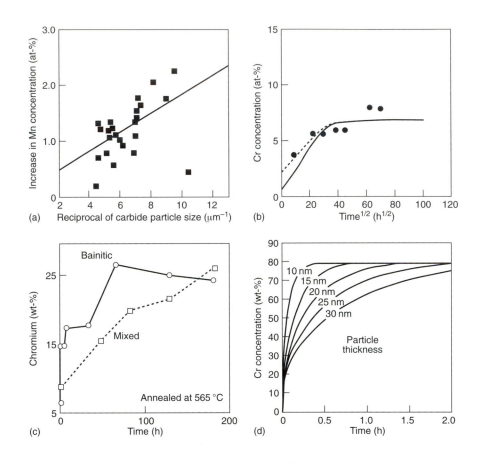

Fig. 8 (a) Size dependence of cementite enrichment (annealed at 565°C for 4 weeks).[41] (b) Data for the enrichment of cementite associated with pearlite, and calculated curves assuming zero enrichment at $t = 0$ (continuous curve), and a finite enrichment at $t = 0$ (dotted line).[42] (c) Cementite enrichment in a fully bainitic microstructure and one which is a mixture of allotriomorphic ferrite and bainite.[43,44] (d) Rapid saturation of cementite in a 12Cr1MoV steel aged at 700°C, as calculated for a variety of particle sizes.[42]

The ternary Fe–Cr–C phase diagram on the $(Fe,X)_3C/\alpha$ field shows that an increase in the carbon concentration is accompanied by a decrease in the equilibrium concentration of chromium in the carbide. Thus the carbide enrichment rate is expected to decrease. A further effect is that the volume fraction of cementite increases, causing an increase in particle thickness and volume fraction. The thickness increase retards the rate of enrichment (eqn 8). If the carbide particles are closer to each other then soft-impingement occurs at an earlier stage, giving a slower enrichment rate at the later stages of annealing.

Local variations in carbon concentration may have a similar effect to changes in average concentration. Such variations can be present through solidification induced segregation, or because of microstructure variations caused by differences in cooling rates in thick sections. It is well known that the microstructure near the component surface can be fully bainitic with the core containing a large amount of allotriomorphic ferrite in addition to bainite. In the latter case, the bainite which grows after the allotriomorphic ferrite, transforms from high carbon austenite. The associated carbides are then found to enrich at a slower rate (Fig. 8c). This discussion emphasises the role of carbon, an element which unfortunately is not tightly specified in creep resistant alloys.

COMPOSITION CHANGES IN ALLOY CARBIDES

The vast majority of steels used in the power generation industry have a total alloy concentration less than 5 wt.%, but there are richer martensitic alloys such as the 9Cr1Mo, 12Cr1MoV destined for more stringent operating conditions. For low alloy steels, the initial carbide phase found immediately after the stress-relief heat treatment (typically several hours at 700°C) is cementite. The cementite is very slow to enrich and to convert to alloy carbides, thus allowing the process to be followed over a period of many years during service. As the concentration of carbide forming elements is raised, the kinetics of cementite enrichment and alloy carbide precipitation become more rapid (Fig. 8d). Thus, relatively stable alloy carbides tend to dominate the microstructure after short periods in service or even after the stress relief heat treatment.[42] The alloy carbides form with virtually their equilibrium compositions, and hence show no change on annealing. There is therefore little prospect of estimating the thermal history of high alloy steels by following changes in carbide chemistries.[42]

THEORY FOR NON-EQUILIBRIUM GROWTH

We have discussed, in some detail, the experimental evidence for the non-equilibrium composition of some carbides when they first form. Non-equilibrium growth is of utmost importance in any description of carbide precipitation in steels. This can be illustrated by examining the limiting case, where the structural transformation to the carbide involves only the diffusion of carbon. An isothermal section of a calculated ternary phase diagram of the Fe–Cr–C system is shown in Fig. 9. The domains in which different carbide phases, or combinations of phases are illustrated in Fig. 10. The lines connecting different phase fields are called *tie-lines*; their end-points represent the chemical compositions of the phases which are in equilibrium with each other. Note that the tie-lines are either measured experimentally or calculated using thermodynamic data – they are not in general lines which radiate from the corners of the plot.

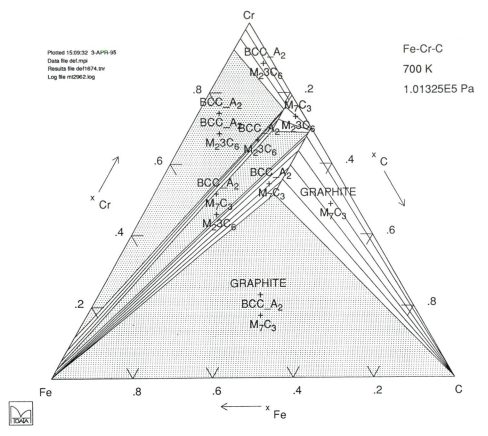

Fig. 9 An isothermal section (700 K) of the Fe–Cr–C equilibrium phase diagram. The BCC phase is ferrite, and *M* stands for a mixture of iron and chromium atoms.

By contrast, Fig. 10 shows a similar phase diagram, but one which represents the paraequilibrium state. Two phases which are in paraequilibrium must have the same ratio of iron to substitutional-solute atoms. Therefore, all the tie-lines must radiate from the carbon corner of the plot, since such lines represent the locus of all points along which the Fe/Cr ratio is constant. The carbon concentration of course varies along each of these tie-lines, being maximum at the carbon corner and minimum at the other end. A comparison of the equilibrium and paraequilibrium versions shows that the change in the form of the diagram when chromium does not partition is quite spectacular.

The paraequilibrium diagram shows that there are remarkable changes in the thermodynamics of the system when there are deviations from equilibrium. Paraequilibrium is in this sense an extreme case for substitutional solutes, which do not partition at all.

Robson has done some interesting calculations which show the importance

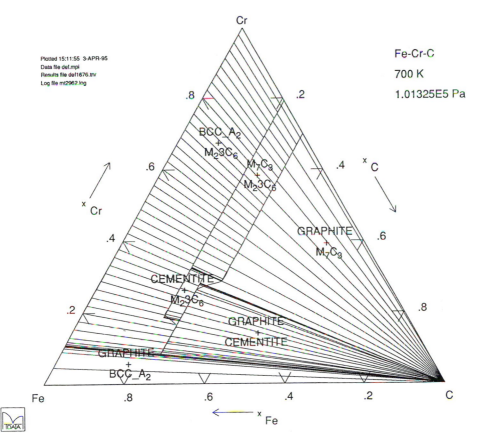

Fig. 10 An isothermal section (700 K) of the Fe–Cr–C *paraequilibrium* phase diagram. The BCC phase is ferrite, and *M* stands for a mixture of iron and chromium atoms.

of deviations from equilibrium, in determining the sequence of precipitation reactions in commercially well-established power plant alloys.[44] The calculations illustrated in Fig. 11 are for the classical 2.25Cr1Mo steel studied by Baker and Nutting.[46] When tempered at elevated temperatures, beginning with a martensitic or bainitic microstructure, there follows the following (simplified) sequence of carbide precipitation reactions:

$$M_3C \rightarrow M_3C + M_2C \rightarrow M_{23}C_6$$

The cementite that precipitates first is not of the equilibrium composition: it contains far less chromium than is expected from the phase diagram. Fig. 11 shows that M_2C would not form at all over a large range of temperatures, in the presence of equilibrium cementite. 2.25Cr1Mo steels would be far less useful as engineering materials if cementite grew with its equilibrium composition!

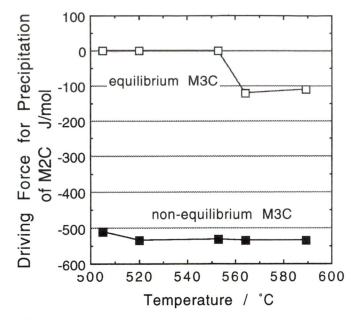

Fig. 11 The calculated driving force for the precipitation of M_2C carbides in a 2.25Cr1Mo steel during elevated temperature tempering. There are two cases illustrated, one in which the first phase to form is equilibrium cementite. M_2C therefore cannot form over a wide range of temperatures where it is observed in practice. In the second case, the cementite forms with a much lower chromium concentration than is expected from equilibrium considerations, making M_2C precipitation thermodynamically possible.

Some of the first data on the chemical composition of cementite in power plant steels were by Baker and Nutting[46] for the 2.25Cr1Mo steel. Their data, measured by extraction and X-ray fluorescence, are illustrated in Fig. 12, as a function of the time and temperature at the tempering temperature. The cementite enriches in Cr, Mo and Mn as diffusion permits, emphasising that it forms initially without equilibrium.

FUTURE WORK: THEORY FOR NON-EQUILIBRIUM GROWTH
We now describe how the composition of the product phase (carbide) may in principle be predicted during its non-equilibrium growth. All current models for predicting the initial cementite composition assume the extreme condition of paraequilibrium; an application of the theory outlined below should help predict the actual (non-equilibrium) composition during growth.

The rate at which an interface moves depends both on its intrinsic mobility (related to the process of structural change across the interface) and on the ease with which any solute elements partitioned during transformation dif-

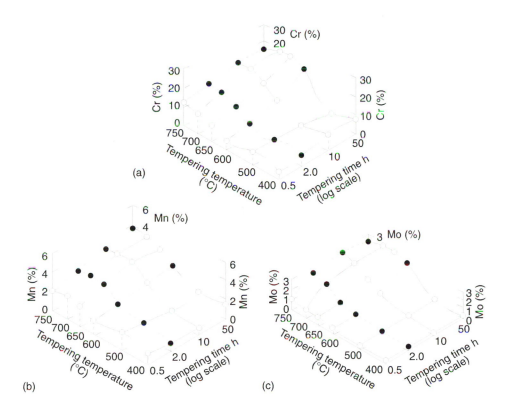

Fig. 12   Diagrams showing the concentrations of Cr, Mn and Mo in extracted carbides as a function of the time and temperature of annealing. The starting microstructure in this case was bainite obtained by a normalising heat-treatment prior to tempering.[46]

fuse ahead of the moving interface. The two processes are in series so that the interfacial velocity as calculated from the interfacial mobility always equals that computed from the diffusion of solute ahead of the interface.

The process of structural change across the interface, and the diffusion of solute ahead of the interface, both dissipate a proportion of the net free energy ΔG that is available for interfacial motion. The two dissipations are G_{id} and G_{dd} respectively, such that

$$\Delta G = G_{id} + G_{dd} \qquad (9)$$

Interface-controlled growth is said to occur when $\Delta G \simeq G_{id}$. Diffusion-controlled growth occurs when $\Delta G \simeq G_{dd}$. Mixed control arises when neither process dominates. Note that interfacial motion is strictly always under mixed control since either process must cause the dissipation of some free energy.

The two ways of calculating the interface velocity are

$$V_i = \psi\{G_{id}\} \tag{10}$$

$$V_d = \psi\{G_{dd}\} \tag{11}$$

where V_i and V_d represent interface-controlled and diffusion-controlled growth rates respectively. Because the interface and diffusion process are in series, substitution of the correct values of G_{dd} and G_{id} into these equations makes $V = V_i = V_d$. This provides a method for fixing G_{id} and G_{dd} for a given value of ΔG, as illustrated in Fig. 14. Thus, for any chosen value of x_θ, the dissipations can be adjusted until $V_i = V_d$, thus fixing the value of x_α. Note that when growth occurs under equilibrium conditions, $x_\alpha = x_{\alpha\theta}$ and $x_\theta = x_{\theta\alpha}$, where $x_{\alpha\theta}$ is the concentration in ferrite which is in equilibrium with the cementite (given by the classical common tangent construction).

Notice that by conducting the analysis just described, it is possible to obtain the interface velocity V as a function of the chromium concentration x_θ in the cementite, but it is not possible to fix the velocity and x_θ that is found in practice. A further condition is needed to fix the velocity and x_θ.

This additional interface response function is provided by Aziz's[47,46] solute trapping law. His model relates interfacial velocity to the partitioning coefficient k_p, which is the ratio of the solute concentration in the (homogeneous) product phase to the solute concentration in the parent phase at the interface. In the present context,

$$k_p = \frac{x^\theta}{x^\alpha} \tag{12}$$

and $k_p = k_e$ when the two concentration terms represent the respective equilibrium concentrations of the phases concerned; k_e is thus the equilibrium partitioning coefficient.

The model is based on chemical reaction rate theory, one of the assumptions of which is that the 'reaction' consists of the repetition of unit steps involving the interaction of a small number of atoms. Aziz gives a velocity function of the form[8-10]

$$V_k = \frac{D}{\lambda}\left(\frac{k_p - k_e}{1 - k_p}\right) \tag{13}$$

where λ is the intersite distance of ≈ 0.25 nm and D is the solute diffusivity in the parent phase.

The methodology outlined here should in principle enable the calculation of the composition of carbides as they grow. This assumes that the interface mobility function is known. And it is also necessary somehow to approxi-

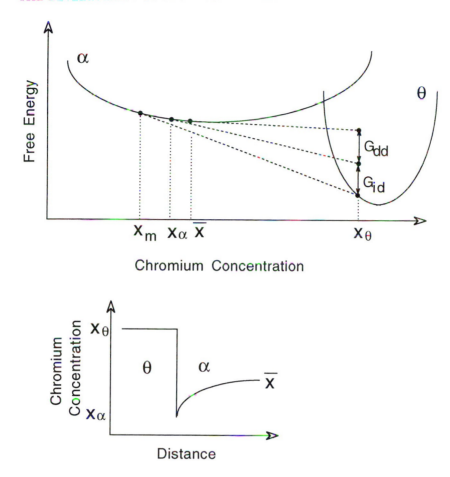

Fig. 13   (a) Constant temperature free energy curves for ferrite (α) and cementite (θ) showing the dissipations due to diffusion and interfacial processes. x is the average chromium concentration in the alloy, x_θ is the concentration in the cementite during its growth, x_m is the minimum chromium concentration in the ferrite, x_α is the concentration in the ferrite at the interface. (b) Distribution of chromium at the cementite/ferrite interface.

mate multicomponent steels, perhaps by identifying the particular element which contributes most to the diffusion flux.

CARBIDE SEQUENCES
The kinetics of solid-state transformations can usually be represented in the form of 'C'-curves on plots of temperature versus time, each curve describing a particular fraction of transformation achieved during isothermal reaction. The C-curve characteristic arises because at high temperatures (i.e. small undercoolings below the equilibrium transformation temperature) the chemical

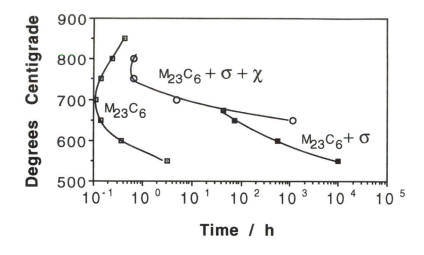

Fig. 14 Time-Temperature-Transformation diagram for precipitation in an austenitic stainless steel (AISI 316), after Weiss and Stickler.[49]

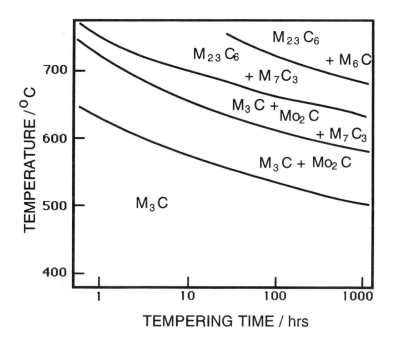

Fig. 15 An adapted form of the Baker-Nutting tempering plot.[46]

driving force for precipitation is small. The driving force increases at lower temperatures but diffusion becomes increasingly difficult as the temperature is reduced. Consequently, there is an optimum undercooling at which the reaction rate is at a maximum. This is the fundamental basis of all time-temperature-transformation (TTT) diagrams; one such diagram for precipitation reactions in an austenitic stainless steel is illustrated in Fig. 14. It shows all the classical features of TTT diagrams, with C-curves.

The same does not seem to apply to precipitation reactions in ferritic steels of the kind used routinely for the construction of power plant. Instead of C-curves, the TTT diagrams appear to show a monotonically increasing rate as the tempering temperature is increased (Fig. 15). A great deal of investigation has revealed that this is because the usual tempering temperatures are far below the equilibrium precipitation temperatures, so that the curves illustrated in Fig. 15 represent the lower half of the expected C-curves.[45]

Aside from the complication outlined above, we have already demonstrated that the carbides formed first in the precipitation reaction can have a profound influence on those formed later in the sequence. A sequence arises in the first place because the equilibrium carbide finds it difficult to nucleate. Metastable carbides which have lower energy interfaces with the matrix form first, even though the reduction of free energy accompanying their precipitation is smaller compared with the precipitation of the equilibrium phase. The prior formation of a metastable phase greatly reduces the chances of forming the equilibrium phase. Thus, it will be seen later that $M_{23}C_6$ precipitates at an incredibly slow rate in 2.25Cr1Mo steel when compared with the 10Cr1Mo type steel. This is because the rapid precipitation of M_2C in the 2.25Cr1Mo alloy greatly reduces the free energy available for the subsequent formation of $M_{23}C_6$.

It can be concluded that precipitation reactions in ferritic power plant steels cannot be modelled by examining individual reactions in isolation; interactions between different precipitates are vital to any model.

The discussion that follows shows how such a complicated problem can be tackled with the help of a new theory for handling simultaneous transformations. It begins with an outline of the well-established theory for individual reactions, followed by an elementary description of the new concepts which are proving to be extremely useful in the physical modelling of power plant steels.

OVERALL TRANSFORMATION KINETICS: ISOLATED REACTIONS

The famous Johnson–Mehl–Avrami theory for overall transformation kinetics is described here. A precipitate particle usually forms after an incubation period τ. Assuming growth at a constant rate G, the volume w_τ of a spherical particle is given by

$$w_\tau = (4\pi/3)G^3(t - \tau)^3 d\tau \ (t > \tau) \tag{14}$$

$$w_\tau = 0 \quad (t < \tau) \tag{15}$$

where t is the time defined to be zero at the instant the sample reaches the isothermal transformation temperature.

Particles nucleated at different locations may eventually touch; this problem of hard impingement is neglected at first, by allowing particles to grow through each other and by permitting nucleation to happen even in regions which have already transformed. The calculated volume of β phase is therefore an *extended volume*; the change in extended volume is given by

$$dV_\beta^e = w_\tau I V d\tau$$

i.e.

$$V_\beta^e = (4\pi V/3) \int_{\tau = 0}^{t} G^3 I(t - \tau)^3 d\tau$$

where $V = V_\gamma + V_\beta$ is the total sample volume. Notice that it is possible to follow the evolution of each extended particle individually.

The actual change in the volume of β must of course be smaller, at

$$dV_\beta = w_\tau I V_\gamma d\tau = w_\tau I(V - V_\beta) \, d\tau$$

It follows that

$$dV_\beta = \left(1 - \frac{V_\beta}{V}\right) dV_\beta^e$$

$$V_\beta^e = - V \ln\left(1 - \frac{V_\beta}{V}\right)$$

so that

$$- \ln\left(1 - \frac{V_\beta}{V}\right) = (4\pi/3)G^3 \int_0^t I(t - \tau)^3 d\tau$$

In making this conversion from extended to real volume, all information about individual particles is lost, so that the application of the Avrami model can only yield average quantities such as grain size, volume fraction but not grain size or grain volume distributions.

The equation can be integrated with specific assumptions about the nucleation rate – when the nucleation rate is constant:

$$\frac{V_\beta}{V} = 1 - \exp(-\pi G^3 I t^4/3)$$

This is the form of the Avrami equation, which can be adapted for a variety of nucleation and growth phenomena and indeed, for highly complex phenomena such as the overlap of the diffusion fields of adjacent growing particles. It nicely and quantitatively describes the C-curve behaviour mentioned earlier, but is restricted to the precipitation of a single phase from the parent phase. We now proceed to illustrate how such theory can be modified to treat the occurrence of several simultaneous reactions as in power plant alloys.

NEW KINETIC THEORY FOR SIMULTANEOUS REACTIONS

Consider the precipitation of two phases, α and β from the parent phase γ. For power plant steels, these phases are carbides or intermetallic compounds whose volume fractions are very small even at equilibrium, so that the conversion from extended to real volume can be neglected. It follows from the previous procedure, that the changes in the volumes of α and β are given by:

$$dV_\alpha = \frac{4}{3} \pi G_\alpha^3 (t - \tau)^3 I_\alpha (V - V_\alpha - V_\beta) \, d\tau \tag{16}$$

$$dV_\beta = \frac{4}{3} \pi G_\beta^3 (t - \tau)^3 I_\beta (V - V_\alpha - V_\beta) \, d\tau \tag{17}$$

assuming spherical particles which grow at constant rate without soft-impingement. These equations can no longer be analytically integrated because the quantities V_a and V_β are functions of each other. The procedure is therefore as follows:

(i)   The initial volumes of α and β are set to zero.
(ii)   The two equations are placed in a pair of nested 'do-loops', the inner loop allowing increments in τ, the outer for increments in t.

After the first inner loop is completed, and on each successive completion, the values of V_α and V_β are updated, allowing for any changes in the growth and nucleation rates as well, due to potential changes in the chemical composition of the remaining parent phase. Both dissolution and growth can therefore be handled, e.g. dissolution of cementite to permit the growth of M_2C.

For large volume fractions of transformation, the relationship between the change in extended and real volumes can be obtained by comparing equations 18 and 19 with 16 and 17:

$$dV^e_\alpha = \frac{4}{3} \pi G^3_\alpha (t - \tau)^3 I_\alpha V \, d\tau \tag{18}$$

$$dV^e{}_\beta = \frac{4}{3}\, \pi\, G^3{}_\beta (t - \tau)^3 I_\beta V \, d\tau \qquad (19)\dagger$$

Some example calculations are presented in Fig. 16. When the nucleation and growth rates of α and β are set to be exactly identical, their curves for volume fraction versus time are exactly superimposed and each phase eventually achieves a fraction of 0.5 (Fig. 16a). When the nucleation rate of β is set to equal twice that of α, then for identical growth rates, the terminal fraction of β is, as expected, twice that of α (Fig. 16b). The case where the growth rate of β is set to be twice that of α (with identical nucleation rates) is illustrated in Fig. 16c.

A more complicated description for a power plant steel is illustrated in Fig. 17. The cementite forms first, begins to dissolve with the precipitation of both M_2C and $M_{23}C_6$, with M_2C beginning to dissolve at much longer times. It is interesting to compare this alloy (a 10Cr1Mo steel) with the Baker and Nutting (2.25Cr1Mo steel). $M_{23}C_6$ precipitation is very slow indeed in the Baker and Nutting steel, because the relatively rapid precipitation of M_2C stifles that of $M_{23}C_6$. The fraction of M_2X is smaller in the 10Cr1Mo steel so that $M_{23}C_6$ precipitates rapidly. These phenomena could not have been predicted without the treatment of simultaneous kinetics, and provide a method for delaying the precipitation of $M_{23}C_6$.

MARTENSITE

The 12Cr and 9Cr steels are mostly martensitic. Their martensite-start (M_s) temperature is important in defining the welding procedure, and empirical equations are known to be inaccurate in predicting M_s. Whatever the details of the mechanism of martensitic nucleation, one very important result is that martensite is triggered when the temperature T is such that free energy change for the diffusionless transformation of the parent phase γ to the product phase α' i.e. $\Delta G^{\gamma \to \alpha'}$ becomes less than a critical value $\Delta G^{\gamma \to \alpha'}_{Ms}$

(Fig. 18). This critical value is about -1100 J mol^{-1} for low-alloy steels. However, Ghosh and Olson [51, 52] have shown that it depends on the strength of the parent phase. Dissolved elements which strengthen the austenite make martensitic transformation more difficult and this effect alone can depress the M_s temperature. Thus, the value of $\Delta G^{\gamma \to \alpha'}_{Ms}$ may be much smaller (say

-3000 J mol^{-1} for steels which have large concentrations of strengthening elements. Consequently, it is now possible to estimate the M_s temperature for

† These equations are rigorous for the circumstances described, but there are some difficulties when, in a diffusion-controlled reaction, the solubility of solute in the matrix phase depends on whether the matrix is in equilibrium with α or β. [50]

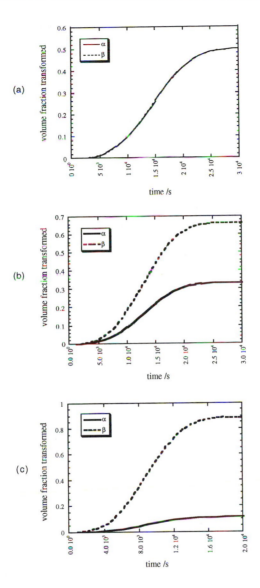

Fig. 16 An illustration of the kinetics of two reactions occurring simultaneously. (a) When the two phases have identical nucleation and growth rates. (b) Identical growth rates but with the β having twice the nucleation rate of α. (c) Identical nucleation rates but with β particles growing at twice the rate of the α particles.

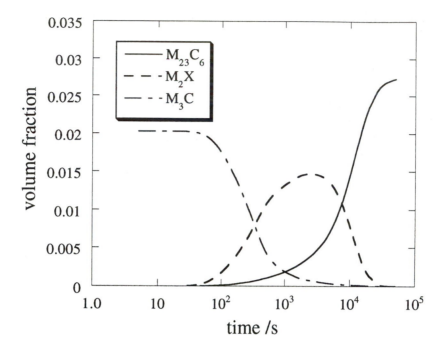

Fig. 17 Calculations for a 10Cr1Mo steel, Fe–0.11C–0.5Mn–10.22Cr–1.42Mo–0.55Ni–0.2V–0.5Nb wt.% steel, heat treated at 600°C.

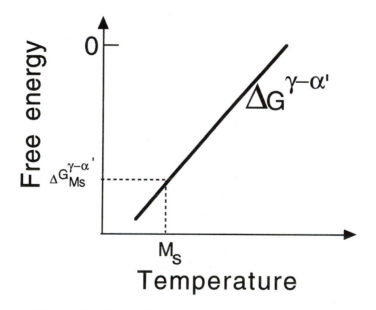

Fig. 18 Martensite forms when the driving force for diffusionless transformation achieves a critical value.

the whole range of power plant steels to an accuracy of $\pm 20°C$ as a function of alloy chemistry.

CONCLUSIONS

An attempt has been made to show that the methods available for modelling the development of microstructure in power plant alloys are making reasonable progress. The work has been illustrated using selected examples – many important developments have not been introduced due to a lack of space. These include the problem of the redistribution of carbon during the heat treatment of dissimilar steel joints, and the modelling of mechanical properties, both of which are well-developed subjects in the context of mathematical modelling.

ACKNOWLEDGMENTS

The authors are grateful to National Power plc. for financial support via Dr. David Gooch. We would also like to thank David Gooch, Andrew Strang, Rod Vanstone and Rachel Thomson for their help during the course of some of this work. The help of Hugh Davies and Susan Hodgson with the provision and support of MTDATA is greatly appreciated.

REFERENCES

1. K.J. Irvine: *Materials in Power Plant*, Institution of Metallurgists, London, 1975, 1–10.
2. I.M. Austin, P. McIntyre and E.F. Walker: *Materials in Power Plant*, Institution of Metallurgists, London, 1975, 120–127.
3. D. Radaj: *Mathematical Modelling of Weld Phenomena 2*, H. Cerjak and H.K.D.H. Bhadeshia eds, Institute of Materials, London 1995, 245–264.
4. M. Rothman: *An Introduction to Industrial Mathematics*, G.T. Foulis & Co. Ltd., Oxfordshire, 1970, 198–207.
5. *Neural Network PC Tools*, R.C. Eberhart and R.W. Dobbins eds, Academic Press, New York, 1990.
6. A.J. Summerfield, T. Cool, B. Keville and H.K.D.H. Bhadeshia: *Welding and Metal Fabrication*, 1995, **64**, 23–24.
7. R.W.K. Honeycombe and H.K.D.H. Bhadeshia: *Steels, Microstructure and Properties*, Edward Arnolds, 2nd edition, London, 1995, Chapter 14.
8. M. Kranzberg and C.S. Smith: *Materials Science and Engineering*, 1979, **37**, 1–40.
9. A.B. Pippard, *The Elements of Classical Thermodynamics*. Cambridge University Press, Cambridge, 1957, Chapter 1.
10. O. Kubaschewski and E. Ll. Evans: *Metallurgical Thermochemistry*, Pergamon Press, Oxford, 1950.
11. T.G. Chart, J.F. Counsell, G.P. Jones, W. Slough and J.P. Spencer: *International Metals Reviews*, 1975, **20**, 57–82.

12. L. Kaufman: *Progress in Materials Science*, 1969, **14**, 57–96.

13. M. Hillert: *Hardenability Concepts with Applications to Steel*, D.V. Doane and J.S. Kirkaldy eds, TMS–AIME, Warrendale, PA, 1977, 5–27.

14. *The SGTE Casebook: Thermodynamics at Work*, K. Hack ed., Materials Modelling Series, Institute of Materials, London, 1996.

15. MTDATA: *Metallurgical and Thermochemical Databank*, National Physical Laboratory, Teddington, Middlesex, UK, 1995.

16. W.C. Leslie: *Metallurgical Transactions*, 1972, 5–26.

17. S. Takeuchi: *Journal of the Physical Society of Japan*, 1969, **27**, 929–940.

18. L.-E. Svensson: *Control of Microstructure and Properties in Steel Arc Welds*, CRC Press, London, 1994.

19. J.W. Gibbs: *The Scientific Papers of J.W. Gibbs*, Dover Publications, New York, 1961, 57.

20. L.S. Darken: *TMS-AIME*, 1949, **180**, 430.

21. L.S. Darken and R.W. Gurry: *Physical Chemistry of Metals*, McGraw-Hill, New York, 1953, 146–148.

22. J.C. Baker and J.W. Cahn: *Solidification*, ASM, Metals Park, Ohio, 1971, 23.

23. J.W. Cahn: *Bull. Alloy Phase Diagrams*, 1980, **1**, 27.

24. A. Hultgren: *Jernkontorets Ann.*, 1951, **135**, 403.

25. M. Hillert: *Jernkontorets Ann.*, 1952, **136**, 25–37.

26. E. Rudberg: *Jernkontorets Ann.*, 1952, **136**, 91.

27. J.B. Gilmour, G.R. Purdy and J.S. Kirkaldy: *Metallurgical Transactions*, 1972, **3**, 1455.

28. M. Enomoto and H.I. Aaronson: *Scripta Metallurgica*, 1985, **19**, 1–3.

29. H.K.D.H. Bhadeshia: *Mathematical Modelling of Weld Phenomena 2*, H. Cerjak ed., Institute of Materials, London, 1995, 71–118.

30. H.K.D.H. Bhadeshia: *Materials Science and Technology*, 1989, **5**, 131–137.

31. S.V. Tsivinsky, L.I. Kogan and R.I. Entin: *Problems of Metallography and the Physics of Metals*, B. Ya Lybubov ed., State Scientific Press, Moscow (Translation published by the Consultants Bureau Inc., New York, 1959), 1955, 1985–1991.

32. J. Chance and N. Ridley: *Metallurgical Transactions A*, 1981, **21A**, 1205–1213.

33. K.W. Andrews: *Acta Metallurgica*, 1963, **11**, 939–946.

34. W. Hume-Rothery, G.V. Raynor and A.T. Little: *Arch Eisenhüttenwesen*, 1942, **145**, 143.

35. H.C. Yakel: *International Metals Reviews*, 1985, **30**, 17–40.

36. P.J. Sandvik: *Metallurgical Transactions A*, 1982, **13A**, 789–800.

37. T. Nakamura and S. Nagakura: *International Conf. on Martensitic*

Transformations, ICOMAT'86, The Japan Institute of Metals, Tokyo, 1986, 1057–1065.

38. K.A. Taylor, G.B. Olson, M. Cohen and J.B. Vander Sande: *Metallurgical Transactions A*, 1989, **20A**, 2749–2765.

39. K.A. Taylor, L. Chang, G.B. Olson, G.D.W. Smith, M. Cohen and J.B. Vander Sande: *Metallurgical Transactions A*, 1989, **20A**, 2772–2737.

40. H.K.D.H. Bhadeshia: *Materials Science and Technology*, 1989, **5**, 131–137.

41. P. Wilson: *Ph.D. Thesis*, University of Cambridge, Cambridge, UK, 1991.

42. R.C. Thomson and H.K.D.H. Bhadeshia: *Metallurgical Transactions A*, 1992, **23A**, 1171–1179.

43. R.C. Thomson and H.K.D.H. Bhadeshia: *Materials Science and Technology*, 1994, **10**, 193–204.

44. R.C. Thomson and H.K.D.H. Bhadeshia: *Materials Science and Technology*, 1994, **10**, 205–208.

45. J.D. Robson: *Unpublished research*, University of Cambridge, UK, 1995.

46. R.G. Baker and J. Nutting: *Journal of the Iron and Steel Institute*, 1959, **192**, 257–268.

47. M.J. Aziz: *J. Applied Physics*, 1982, **53**, 1158–1168.

48. M.J. Aziz: *Applied Physics Letters*, 1983, **43**, 552–554.

49. B. Weiss and R. Stickler: *Metall. Trans.*, 1972, **3**, 851.

50. J.D. Robson and H.K.D.H. Bhadeshia: *Unpublished research*, University of Cambridge, UK, 1995.

51. G. Ghosh and G.B. Olson: *Acta Metall. et Mater.*, 1994, **42**, 3361–3379.

52. T. Cool and H.K.D.H Bhadeshia: *Materials Science and Technology*, 1996, **12**, 40–44.

List of Delegates

Auerkari, P.;
VTT, Finland

Artinger, I.;
TU Budapest, Hungary

Barnes, Adrienne;
TWI, UK

Beech, S.M.;
Rolls-Royce IRD, UK

Berger, Christina;
Technische Hochshule Darmstadt,
Germany

Bhadeshia, H.K.D.H.; (179–207)
Cambridge University, UK

Bianchi, Mrs P.;
ENEL-CRAM, Italy

Bjarbo, A.O.;
Royal Inst. Technology, Sweden

Brandle, M.;
Siemens KWU, Germany

Browne, B.;
PowerGen, UK

Cawley, J.;
Sheffield Hallam University, UK

Cerjak, H.; (93–104, 105–122, 145–158)
TU Graz, Austria

Cool, Tracey;
Parsons/Cambridge University, UK

Corti, S.;
CISE, Italy

Ennis, P.J.;
Research Centre Juelich, Germany

Evans, G.;
Perlikon, Switzerland

Feng, G.;
Westlinghouse, USA

Foldyna, V.; (73–92, 93–104, 145–158)
Vitkovice Research Centre,
Ostrava, Czech Republic

Gass, D.;
Parsons, UK

Gooch, D.J.;
National Power, UK

Haigh, Rachel;
Birmingham University, UK

Hald, J.; (93–104, 123–144, 159–178)
ELSAM/Elkraft, Denmark

Hamerton, R.G.;
AEA Technology, UK

Hanus, R.;
Voest Alpine, Austria

Henderson, M.;
DRA Pyestock, UK

Hofer, P.; (105–122, 145–158)
TU Graz, Austria

Honeyman, G.;
Forgemasters Engineering, UK

Jakobova, Anna; (73–92)
Vitkovice Research Centre,
Ostrava, Czech Republic

Johansson, L.;
ABB Stahl AB, Sweden

Kadoya, Y.;
Imperial College, UK

Kasl, J.;
Skoda, Czech Republic

Kern, T.-I.;
Siemens KWU, Germany

Keville, Barbara;
Oberlikon, Switzerland

Lee, M.;
National Power, UK

Lundin, L.M.;
Chalmers University, Sweden

Mayer, K.-H.; (105–122)
MAN Energie GmbH, Germany

Metcalfe, B.;
Special Melted Products, UK

Metcalfe, E.; (123–144)
National Power, UK

Authors in these proceedings have the corresponding page numbers listed after their names.

Montagnon, J.;
Aubert et Duval, France
Nath, B.; (123–144)
National Power, UK
Nishimura, N.;
Imperial College, UK
Orr, J.; (53–72)
British Steel, UK
Pickering, F.B.; (1–30)
Consultant, UK
Potthast, E.;
Saarschmiede, Germany
Race, Julia;
Parsons, UK
Robson, J.;
Cambridge University, UK
Scarlin, R.B.;
ABB, Switzerland
Schaff, H.;
Aubert et Duval, France
Schonfield, K.-H.;
Saarschmiede, Germany
Schuster, F.; (105–122)
Voest Alpine, Austria
Shaw, N.;
National Power, UK
Silitonen, P.T.;
Tampere University, Finland
Sklenicka, V.;
Inst. of Physics, Brno, Czech
Republic

Spiradek, Krystyna; (93–104)
Austrian Research Centre, Austria
Staubli, M.;
ABB, Switzerland
Stiens, R.;
Inst. of Technology, Zurich,
Switzerland
Stief, J.;
PHB Stahlguss GmbH, Germany
Strang, A.; (31–52)
GECA, UK
Thornton, D.V.;
GECA, UK
Vanecek, V.;
Skoda, Czech Republic
Vanstone, R.; (93–104)
GECA, UK
Vodarek, V.; (31–52, 73–92)
Vitkovice/GECA, Czech Republic
Whitton, A.;
Aubert et Duval, France
Williams, J.A.;
Consultant, UK
Woollard, Louisa; (53–72)
British Steel, UK
Zeiler, G.;
Bohler, Austria

Index